Canadian Mathematical Society
Société mathématique du Canada

More information about this series at http://www.springer.com/series/4318

Ralf Schiffler

Quiver Representations

Ralf Schiffler
Department of Mathematics
University of Connecticut
Storrs, CT, USA

ISSN 1613-5237 ISSN 2197-4152 (electronic)
ISBN 978-3-319-36317-2 ISBN 978-3-319-09204-1 (eBook)
DOI 10.1007/978-3-319-09204-1
Springer Cham Heidelberg New York Dordrecht London

Mathematics Subject Classification (2010): 16S70; 16G20

Printed on acid-free paper

Springer is part of Springer Science+Business Media (www.springer.com)

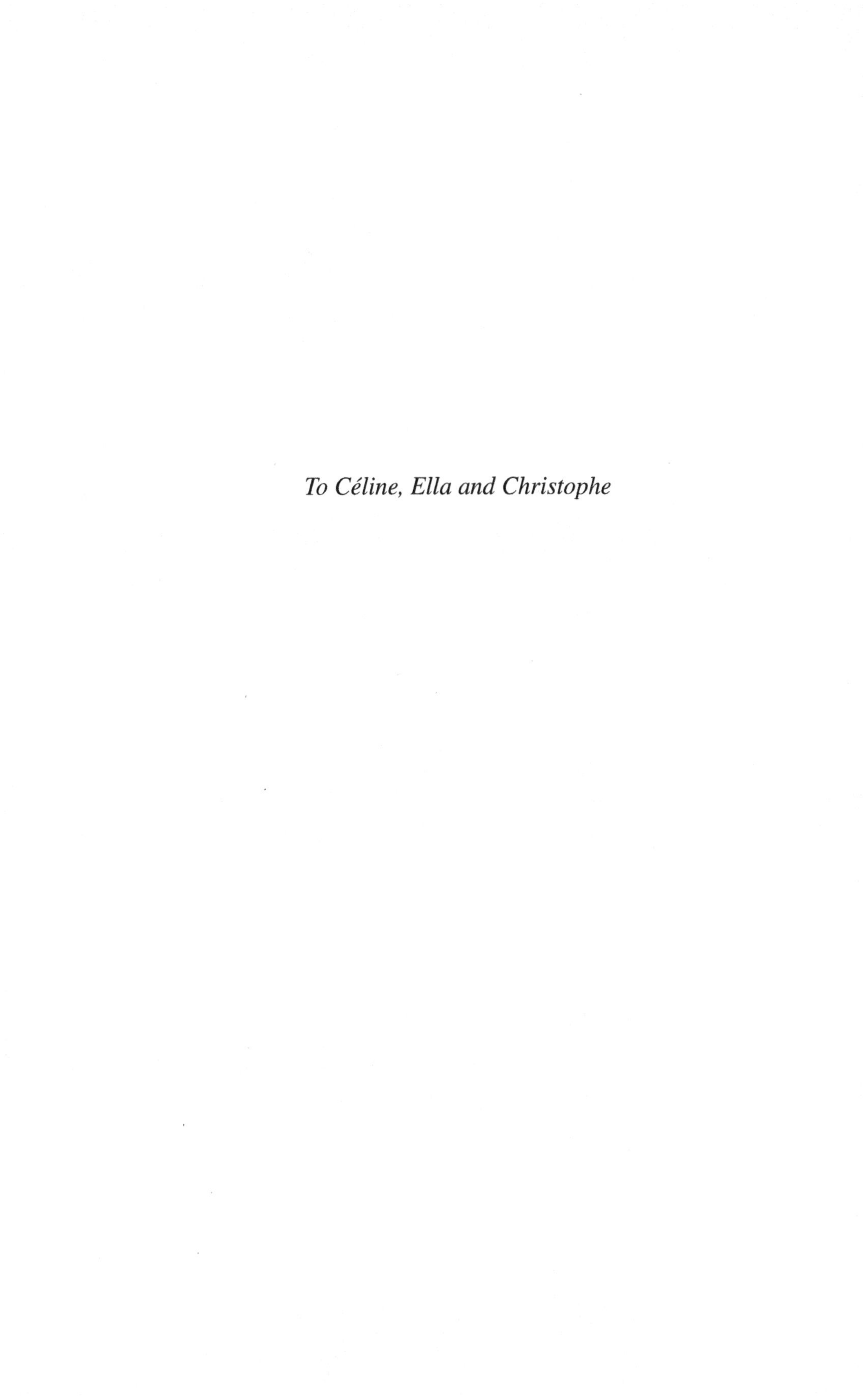

To Céline, Ella and Christophe

Preface

This textbook is an introduction to the representation theory of finite-dimensional algebras with a strong emphasis on quivers and their representations. The book is intended to be used as a graduate textbook for a one-semester course. The first three chapters are completely self-contained, assuming only familiarity with basic notions of linear algebra, and could be used also for an advanced undergraduate topics course or as a quick introduction to Auslander–Reiten quivers for mathematicians who are mostly interested in applying the theory to other fields of research without necessarily becoming an expert in representation theory. In Chaps. 4–7, prior experience with rings is beneficial, but the main concepts are recalled in Chap. 4.

The use of quivers in the representation theory of finite-dimensional algebras gives us the possibility to visualize the modules of a given algebra very concretely as a collection of matrices, each of which is associated to an arrow in a certain diagram—the quiver. To every quiver one can associate the path algebra, whose elements are finite sums of paths in the quiver and whose multiplication is given as concatenation of paths. The modules of the path algebra correspond precisely to the representations of the quiver. Thus the quiver does give not only an example of an algebra but also a very concrete model for the representation theory of the algebra. The beauty of the theory is that the quiver approach can be used to study the representation theory of an *arbitrary* finite-dimensional algebra!

The main tool for describing the representation theory of a finite-dimensional algebra is the Auslander–Reiten quiver, which gives explicit information about the modules as well as the morphisms between them in a most convenient way. When making the choices on how to develop the material in this book, my main goal was to get to the construction of Auslander–Reiten quivers as soon as possible. This is why, in the first three chapters, I only use the language of quiver representations, postponing the viewpoint of algebras and modules to Chaps. 4–7. For the student, this approach has the advantage of having the wealth of examples of the first three chapters at hand, when studying the somewhat abstract notion of a module.

Chapter 1 starts with the definition of quivers and their representations and then develops the basic tools such as morphisms, direct sums, exact sequences, etc. The concepts of projective and injective representations as well as the Auslander–Reiten translation are introduced in Chap. 2. Chapter 3 contains various methods for the construction of Auslander–Reiten quivers and describes explicitly how to use them to compute morphisms and extensions between representations. Chapter 4 introduces algebras in general and path algebras in particular, while Chap. 5 is devoted to bound quiver algebras, which are quotients of path algebras by admissible ideals. The proof of the equivalence of the notions of modules over the bound quiver algebra and representations of the bound quiver is given in Chap. 5. In Chap. 6, we present several popular constructions of algebras. The Auslander–Reiten formulas are proved in Chap. 7 and Gabriel's Theorem in Chap. 8. Chapter 8 does not use the results of Chaps. 4–7 and could be read right after Chap. 3.

Representation theory is an ideal context to introduce the student to the basic concepts of category theory, and the language of categories is developed along the way as needed.

The starting point for this book was a graduate course I gave at the 2008 summer school of the Atlantic Association for Research in the Mathematical Sciences held at the University of New Brunswick–Fredericton. I thank the AARMS for their invitation and their support during the time of writing. Many thanks to İlke Çanakçı, Lucas David–Roesler, and Benjamin Salisbury for many valuable comments and suggestions on the presentation of the material.

The following diagram shows how the different chapters depend on each other:

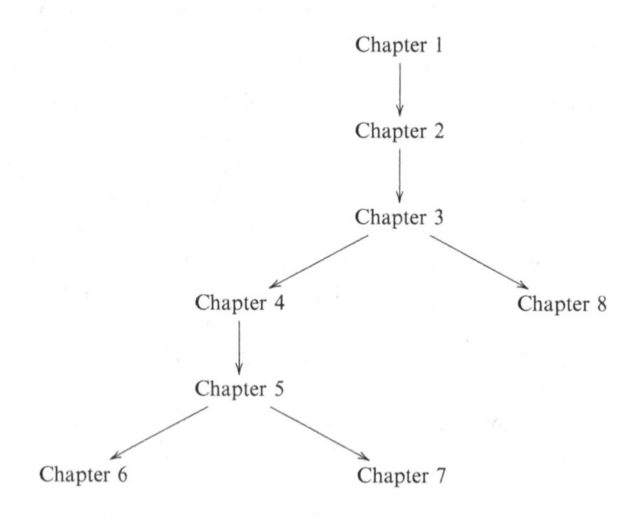

Storrs, CT, USA
September 2013

Ralf Schiffler

Contents

Part I
Quivers and Their Representations

The main idea of representation theory is to study a set with a certain algebraic structure, like a group or an algebra, by looking at its representations. A representation of a group G, for example, is a group homomorphism $\rho : G \to \mathrm{GL}(V)$ from the group G to the group of k-linear isomorphisms from a k-vector space V to itself, where k is some field. Thus each group element $g \in G$ induces an isomorphism $\rho(g) : V \to V$, and, if g and h are two elements of the group G, then $\rho(gh)$ is equal to the composition $\rho(g) \circ \rho(h)$. Hence the group elements are represented via ρ by isomorphisms of the vector space V, and this allows us to use tools from linear algebra to study them.

A different approach to these representations is via the group algebra kG, whose elements are linear combinations of the group elements over the field k and whose multiplication is induced by the group operation. Then the representation theory of the group G is equivalent to the representation theory of its group algebra kG, and in this sense, the representation theory of groups is contained in the representation theory of algebras.

In this book, our interest is the representation theory of finite-dimensional algebras. The representations of a finite-dimensional algebra (over an algebraically closed field) can be described using quivers and their representations.

In this first part of the book, we will work exclusively in the context of quivers and their representations, leaving the algebras and their modules to Part II.

Chapter 1
Representations of Quivers

In this chapter, we introduce the concept of quiver representations and their morphisms, discuss direct sums, kernels, and cokernels, and study short exact sequences of quiver representations. We also introduce some basic notions of category theory.

1.1 Definitions and Examples

A quiver representation is a finite collection of vector spaces and linear maps between these vector spaces. One can visualize this concept using a diagram of arrows, the quiver, where each arrow represents one of the linear maps.

1.1.1 Representations

In order to study quiver representations we need a formal definition of quivers first.

Definition 1.1. A **quiver** $Q = (Q_0, Q_1, s, t)$ consists of

Q_0 a set of vertices,
Q_1 a set of arrows,
$s: Q_1 \to Q_0$ a map from arrows to vertices, mapping an arrow to its starting point,
$t: Q_1 \to Q_0$ a map from arrows to vertices, mapping an arrow to its terminal point.

We will represent an element $\alpha \in Q_1$ by drawing an arrow from its starting point $s(\alpha)$ to its endpoint $t(\alpha)$ as follows:

$$s(\alpha) \xrightarrow{\alpha} t(\alpha).$$

R. Schiffler, *Quiver Representations*, CMS Books in Mathematics,
DOI 10.1007/978-3-319-09204-1_1

Example 1.1. The following quiver is given by $Q_0 = \{1,2,3\}$, $Q_1 = \{\alpha,\beta,\gamma,\lambda,\mu\}$, $s(\alpha) = 3, s(\beta) = 2, s(\gamma) = 3, s(\lambda) = 1, s(\mu) = 1$ and $t(\alpha) = 2, t(\beta) = 1, t(\gamma) = 3, t(\lambda) = 3, t(\mu) = 3$.

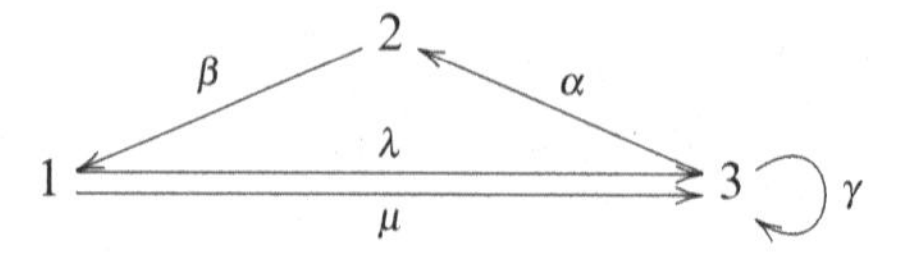

A quiver Q is called *finite* if Q_0 and Q_1 are finite sets. We will always suppose our quivers to be finite.

For the definition of quiver representations, we need a field k. For simplicity, we let k be an algebraically closed field.

Definition 1.2. A representation $M = (M_i, \varphi_\alpha)_{i\in Q_0,\alpha\in Q_1}$ of a quiver Q is a collection of k-vector spaces

$$M_i$$

one for each vertex $i \in Q_0$, and a collection of k-linear maps

$$\varphi_\alpha : M_{s(\alpha)} \to M_{t(\alpha)}$$

one for each arrow $\alpha \in Q_1$.

A representation M is called **finite-dimensional** if each vector space M_i is finite-dimensional. In this case the **dimension vector** $\underline{\dim}\, M$ of M is the vector $(\dim M_i)_{i\in Q_0}$ of the dimensions of the vector spaces. An **element** of a representation M is a tuple $(m_i)_{i\in Q_0}$ with $m_i \in M_i$.

Example 1.2. Let Q be the quiver $1 \to 2$. Then

$$\begin{array}{lll}
M & k \xrightarrow{\;1\;} k \\
M' & k \xrightarrow{\;0\;} k \\
M'' & k \xrightarrow{\;0\;} 0 \\
M''' & k^2 \xrightarrow{\begin{bmatrix}1&0\\1&0\\0&0\end{bmatrix}} k^3
\end{array}$$

are representations of Q. The dimension vectors are $\underline{\dim} M = \underline{\dim} M' = (1, 1)$, $\underline{\dim} M'' = (1, 0)$, and $\underline{\dim} M''' = (2, 3)$.

The subject of this book is to study finite-dimensional quiver representations.

1.1.2 Morphisms

Definition 1.3. Let Q be a quiver and let $M = (M_i, \varphi_\alpha)$, $M' = (M'_i, \varphi'_\alpha)$ be two representations of Q. A **morphism** (or homomorphism) of representations $f: M \to M'$ is a collection $(f_i)_{i \in Q_0}$ of linear maps

$$f_i: M_i \longrightarrow M'_i$$

such that for each arrow $i \xrightarrow{\alpha} j$ in Q_1 the diagram

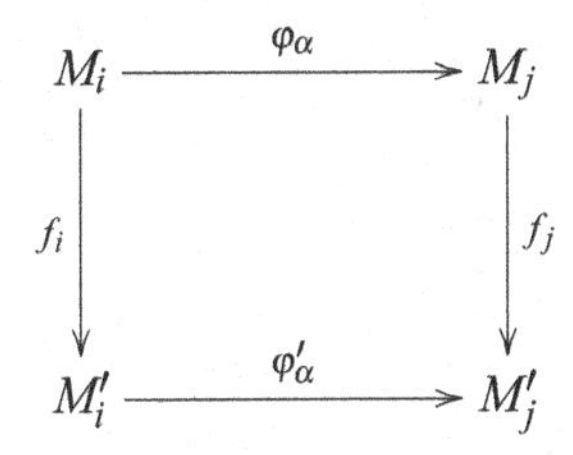

commutes, that is,

$$f_j \circ \varphi_\alpha(m) = \varphi'_\alpha \circ f_i(m) \quad \text{for all } m \in M_i.$$

A morphism $f = (f_i): M \to N$ is an **isomorphism** if each f_i is bijective. The class of all representations that are isomorphic to a given representation M is called the **isoclass** of M.

Example 1.3. Let us consider the representations in Example 1.2 again. The map $f = (f_1, f_2)$, where f_1 is the multiplication by $a \in k$ and f_2 is the zero map, is a morphism from M to M'':

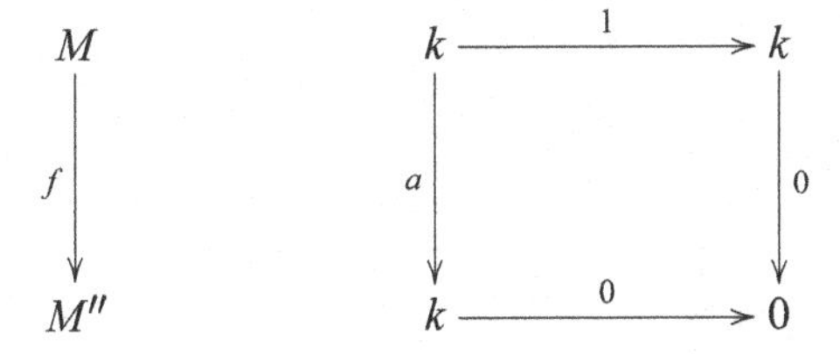

Now let us see if there are there any morphisms $g: M'' \to M$. Suppose we have a commutative diagram:

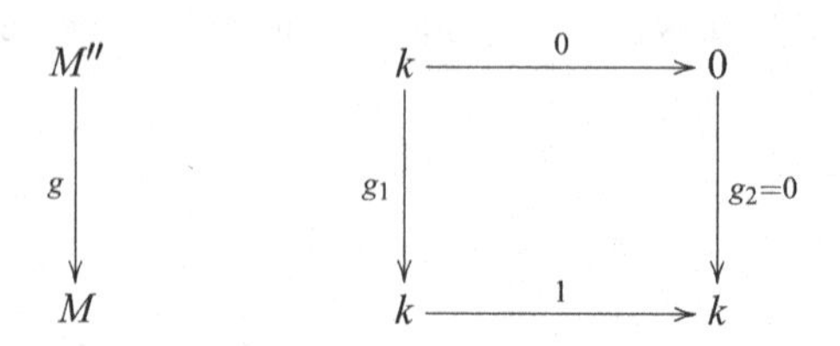

Then $g_1 = 1 \circ g_1$ must be equal to the zero map, and thus $g_1 = 0$. We have shown that the only morphism from M'' to M is the zero morphism $g = (0, 0)$.

Given a quiver Q, the finite-dimensional representations of Q together with the morphisms of representations form a category which we denote by rep Q.

Categories 1 *We will work with categories throughout the book, and we will develop the language of category theory along the way. For a formal definition see Categories 2 at the end of Sect. 1.2. For now, it suffices to know that a category consists of objects and morphisms.*

We write $M \in \text{rep } Q$ if M is an object in rep Q, that is, if M is a finite-dimensional representations of the quiver Q.

Proposition 1.1. *Let $M, M' \in \text{rep } Q$. Then the set of all morphisms $\text{Hom}(M, M')$ is a k-vector space with respect to the addition and scaling of morphisms.*

Proof. Exercise. □

Example 1.4. With the notation of Example 1.3, we have

$$\text{Hom}(M, M'') \cong \{(a, 0) \mid a \in k\} \cong k,$$

where the last isomorphism holds because the vector space $\{(a, 0) \mid a \in k\}$ is of dimension one. On the other hand,

$$\text{Hom}(M'', M) \cong 0.$$

Example 1.5. Let Q be the quiver

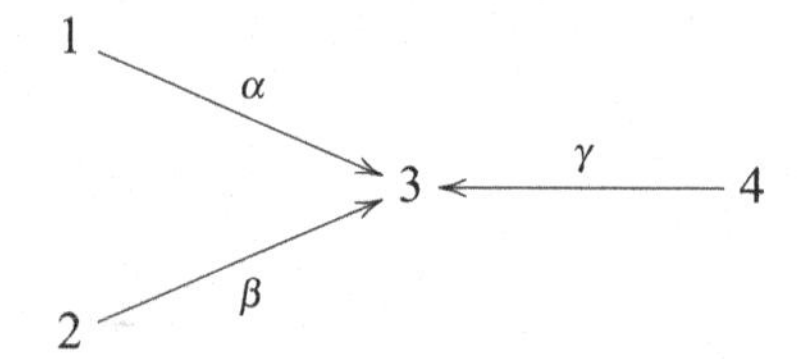

and consider the following representations:

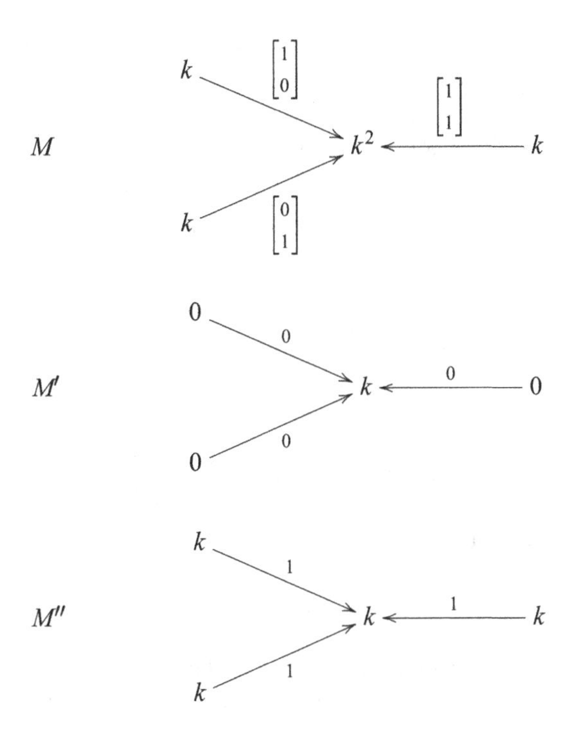

Note that the images of the three maps in M are three different lines in k^2. Then we have

$$\operatorname{Hom}(M, M') = 0 \quad \operatorname{Hom}(M, M'') \cong k^2$$
$$\operatorname{Hom}(M', M) \cong k^2 \quad \operatorname{Hom}(M'', M) = 0.$$

Proof. We show that $\operatorname{Hom}(M, M'') \cong k^2$ and leave the other identities as an exercise. In this example, a morphism $M \to M''$ is a choice of 5 scalars $a, b, c, d, e \in k$ such that the following diagram commutes:

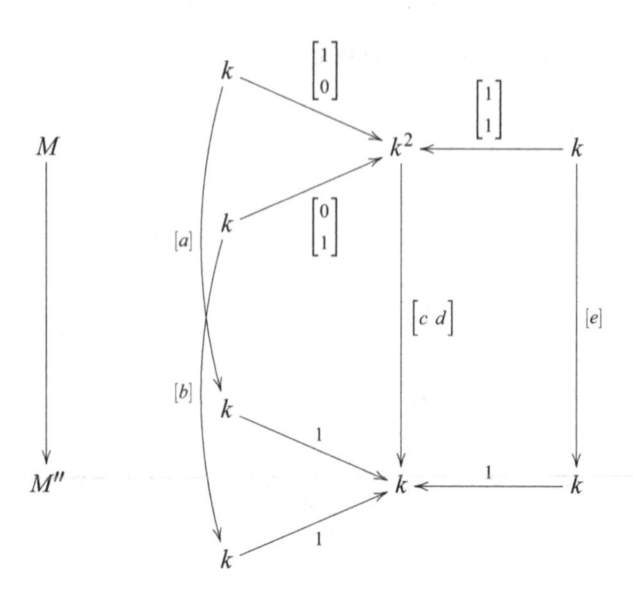

The three commuting squares give the relations

$$a = c, \qquad b = d, \qquad c + d = e.$$

Thus a choice of a and b completely determines the morphism. On the other hand, every choice of a and b yields a different morphism. Therefore $\mathrm{Hom}(M, M'') \cong k^2$. □

Example 1.6. Let Q be the quiver

$$1 \underset{\beta}{\overset{\alpha}{\leftleftarrows}} 2\ ;$$

This quiver is known as the Kronecker quiver.[1] Consider the following representations of Q:

[1]Leopold Kronecker (1823–1891) studied the problem of classifying pairs of matrices of the same size up to simultaneous conjugation, which is equivalent to studying the representations of the Kronecker quiver. The concept of quivers was introduced much later (1972) by Gabriel [33].

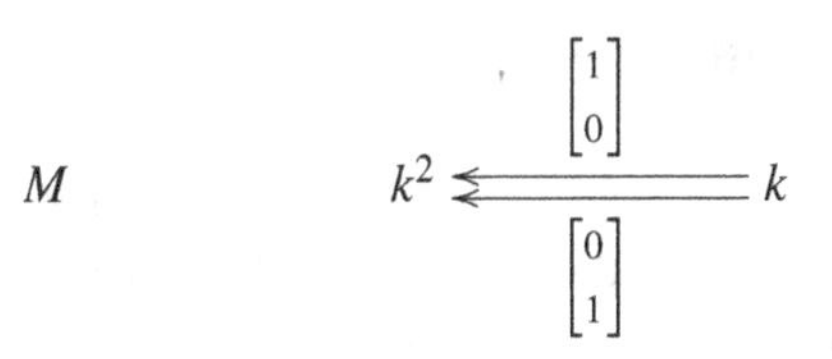

$$M' \qquad k^2 \underset{\begin{bmatrix}0&0\\1&0\end{bmatrix}}{\overset{\begin{bmatrix}1&0\\0&1\end{bmatrix}}{\leftleftarrows}} k^2$$

We want to compute $\mathrm{Hom}(M, M')$. Therefore, suppose that $f = (f_1, f_2)$ is a morphism from M to M'. Then f_1 and f_2 can be written in matrix form as

$$f_1 = \begin{bmatrix} a & b \\ c & d \end{bmatrix} \qquad f_2 = \begin{bmatrix} x \\ y \end{bmatrix}$$

where $a, b, c, d, x, y \in k$, and since f is a morphism of representations, we have $f_1\varphi_\alpha = \varphi'_\alpha f_2$ and $f_1\varphi_\beta = \varphi'_\beta f_2$; in other words

$$\begin{bmatrix} a & b \\ c & d \end{bmatrix}\begin{bmatrix} 1 \\ 0 \end{bmatrix} = \begin{bmatrix} 1 & 0 \\ 0 & 1 \end{bmatrix}\begin{bmatrix} x \\ y \end{bmatrix}, \text{ and}$$

$$\begin{bmatrix} a & b \\ c & d \end{bmatrix}\begin{bmatrix} 0 \\ 1 \end{bmatrix} = \begin{bmatrix} 0 & 0 \\ 1 & 0 \end{bmatrix}\begin{bmatrix} x \\ y \end{bmatrix},$$

which implies that

$$\begin{bmatrix} a \\ c \end{bmatrix} = \begin{bmatrix} x \\ y \end{bmatrix}, \text{ and} \begin{bmatrix} b \\ d \end{bmatrix} = \begin{bmatrix} 0 \\ x \end{bmatrix}.$$

Therefore f is of the form

$$f = \left(\begin{bmatrix} a & 0 \\ c & a \end{bmatrix}, \begin{bmatrix} a \\ c \end{bmatrix}\right),$$

and $\mathrm{Hom}(M, M') \cong k^2$ is a two-dimensional vector space with basis

$$\left\{\left(\begin{bmatrix} 1 & 0 \\ 0 & 1 \end{bmatrix}, \begin{bmatrix} 1 \\ 0 \end{bmatrix}\right), \left(\begin{bmatrix} 0 & 0 \\ 1 & 0 \end{bmatrix}, \begin{bmatrix} 0 \\ 1 \end{bmatrix}\right)\right\}.$$

1.2 Direct Sums and Indecomposable Representations

The direct sum $M \oplus N$ of two representations M and N can be though of as considering both M and N at the same time. If we understand M and N, then we understand their direct sum.

The concept of direct sum is more interesting when we go the other way, that is, given a representation X, we can ask if it is possible to decompose X into a direct sum $X = M \oplus N$, with M and N nonzero. If this is the case, then we can try to decompose the direct summands M and N further and eventually get a decomposition $X = M_1 \oplus M_2 \oplus \cdots \oplus M_t$ in which each of the M_i is indecomposable.

Let Q be a quiver.

Definition 1.4. Let $M = (M_i, \varphi_\alpha)$ and $M' = (M'_i, \varphi'_\alpha)$ be representations of Q. Then

$$M \oplus M' = \left(M_i \oplus M'_i, \begin{bmatrix} \varphi_\alpha & 0 \\ 0 & \varphi'_\alpha \end{bmatrix} \right)_{i \in Q_0, \alpha \in Q_1}$$

is a representation of Q called the **direct sum** of M and M'.

Recursively, we define the direct sum of any finite number of representations $M_1, M_2, \ldots, M_t \in \operatorname{rep} Q$ by

$$M_1 \oplus M_2 \oplus \cdots \oplus M_t = (M_1 \oplus \cdots \oplus M_{t-1}) \oplus M_t.$$

Example 1.7. Let Q be the quiver

$$1 \longrightarrow 2 \longleftarrow 3\,,$$

and consider the representations

$$M \qquad k \xrightarrow{\;1\;} k \xleftarrow{\;0\;} 0;$$

$$M' \qquad k^2 \xrightarrow{\begin{bmatrix} 1 & 1 \\ 0 & 1 \end{bmatrix}} k^2 \xleftarrow{\begin{bmatrix} 1 \\ 1 \end{bmatrix}} k.$$

Then the direct sum $M \oplus M'$ is the representation

$$k \oplus k^2 \xrightarrow{\begin{bmatrix} 1 & 0 & 0 \\ 0 & 1 & 1 \\ 0 & 0 & 1 \end{bmatrix}} k \oplus k^2 \xleftarrow{\begin{bmatrix} 0 & 0 \\ 0 & 1 \\ 0 & 1 \end{bmatrix}} 0 \oplus k;$$

which is isomorphic to

$$k^3 \xrightarrow{\begin{bmatrix} 1&0&0 \\ 0&1&1 \\ 0&0&1 \end{bmatrix}} k^3 \xleftarrow{\begin{bmatrix} 0 \\ 1 \\ 1 \end{bmatrix}} k.$$

Definition 1.5. A representation $M \in \text{rep } Q$ is called **indecomposable** if $M \neq 0$ and M cannot be written as a direct sum of two nonzero representations, that is, whenever $M \cong N \oplus L$ with $N, L \in \text{rep } Q$, then $N = 0$ or $L = 0$.

Example 1.8. The representations in Examples 1.5 and 1.6 are indecomposable. The representation M in Example 1.7 is indecomposable, but M' is not. M' is isomorphic (but not equal) to

$$(k \xrightarrow{1} k \xleftarrow{1} k) \oplus (k \xrightarrow{1} k \xleftarrow{0} 0).$$

In Example 1.2, the representations M and M'' are indecomposable, and the representations M' and M''' are not.

Goal of Representation Theory
Classify all representations of a given quiver Q and all morphisms between them up to isomorphism.

The following theorem shows that in order to attain this goal, it is sufficient to classify all *indecomposable* representations and morphisms between them.

Theorem 1.2 (Krull–Schmidt Theorem). *Let Q be a quiver and let $M \in \text{rep } Q$. Then*

$$M \cong M_1 \oplus M_2 \oplus \cdots \oplus M_t$$

where the $M_i \in \text{rep } Q$ are indecomposable and unique up to order.

Proof. If M is indecomposable, there is nothing to show. If M is not indecomposable, then $M = M' \oplus M''$, where M' and M'' are representations of strictly smaller dimension. By induction, we have $M' \cong M_1' \oplus M_2' \oplus \cdots \oplus M_{t'}'$ and $M'' \cong M_1'' \oplus M_2'' \oplus \cdots \oplus M_{t''}''$ with all M_i', M_i'' indecomposable. This shows the existence of the decomposition. For the uniqueness see, for example, [8, I.4.10]. □

We close this section with the definition of a category.

Categories 2 *A* ***category*** $\mathscr{C}$ *consists of objects, morphisms, and a binary operation called the composition of morphisms.*

More precisely, let $\mathscr{C}$ *be a class of objects* $Ob(\mathscr{C})$ *and a class of morphisms* $Hom_{\mathscr{C}}$ *such that each morphism* $f \in Hom_{\mathscr{C}}$ *has a unique* source X *and a unique* target Y *in* $Ob(\mathscr{C})$. *We say that* f *is a morphism from* X *to* Y *and write* $f : X \to Y$. *The class of all morphisms from* X *to* Y *is denoted by* $Hom_{\mathscr{C}}(X, Y)$.

Then $\mathscr{C}$ is called a *category* if for every three objects X, Y, Z in $\mathrm{Ob}(\mathscr{C})$, there is a binary operation

$$\begin{array}{ccc} \mathrm{Hom}_{\mathscr{C}}(X,Y) \times \mathrm{Hom}_{\mathscr{C}}(Y,Z) & \longrightarrow & \mathrm{Hom}_{\mathscr{C}}(X,Z) \\ (\quad f \quad , \quad g \quad) & \longmapsto & g \circ f \end{array}$$

called the composition of morphisms that satisfies the following axioms:

1. (associativity) If $f : W \to X$, $g : X \to Y$ and $h : Y \to Z$ are morphisms, then
$$h \circ (g \circ f) = (h \circ g) \circ f.$$
2. (identity) For every object X there exists a morphism $1_X \in \mathrm{Hom}_{\mathscr{C}}(X, X)$ called the identity morphism on X such that for every $f \in \mathrm{Hom}_{\mathscr{C}}(X, Y)$ and every $g \in \mathrm{Hom}_{\mathscr{C}}(Z, X)$ we have
$$f \circ 1_X = f \quad \text{and} \quad 1_X \circ g = g.$$

1.3 Kernels, Cokernels, and Exact Sequences

Recall from linear algebra that if $f : V \to V'$ is a linear map, then its kernel $\ker f = \{v \in V \mid f(v) = 0\}$ is a subspace of V, and its cokernel $\operatorname{coker} f = V'/\operatorname{im} f = \{v' + f(V) \mid v' \in V'\}$ is a quotient space of V'.

In this section, we will generalize these concepts to representations.

Let Q be a quiver, and let $M = (M_i, \varphi_\alpha)_{i \in Q_0, \alpha \in Q_1}$ and $M' = (M_i', \varphi_\alpha')_{i \in Q_0, \alpha \in Q_1}$ be two representations of Q. Furthermore, let $f = (f_i)_{i \in Q_0} : M \to M'$ be a morphism of representations. Recall that each f_i is a linear map from the vector space M_i to the vector space M_i'.

For each vertex $i \in Q_0$, let $L_i = \ker f_i$, and for each arrow $i \xrightarrow{\alpha} j$ in Q_1, let $\psi_\alpha : L_i \to L_j$ be the restriction of φ_α to L_i, that is, $\psi_\alpha(x) = \varphi_\alpha(x)$ for all $x \in L_i$. Let us check that ψ_α is well defined. We must show that for all $x \in L_i$, we have $\psi_\alpha(x) \in L_j$ which means that $\varphi_\alpha(x) \in \ker f_j$. But since f is a morphism of representations, we have $f_j \varphi_\alpha(x) = \varphi_\alpha' f_i(x)$, which is zero, since $x \in \ker f_i$. This shows that ψ_α is well defined.

Definition 1.6. The representation $\ker f = (L_i, \psi_\alpha)_{i \in Q_0, \alpha \in Q_1}$ is called the **kernel** of f.

Remark 1.3. The inclusions $\mathrm{incl}_i : \ker f_i \hookrightarrow M_i$ induce an injective morphism of representations[2]:

$$(\mathrm{incl}_i)_{i \in Q_0} : \ker f \hookrightarrow M.$$

Next we define the cokernel of the morphism f. For each vertex $i \in Q_0$, let $N_i = \mathrm{coker}\, f_i = M'_i / f_i(M_i)$, and for each arrow $i \xrightarrow{\alpha} j$ in Q_1, define $\chi_\alpha : N_i \to N_j$ by

$$\chi_\alpha(m'_i + f_i(M_i)) = \varphi'_\alpha(m'_i) + f_j(M_j),$$

for each $m'_i \in M'_i$.

Let us check that χ_α is well defined. Suppose we have two elements $m'_i, m''_i \in M'_i$ such that $m'_i + f_i(M_i) = m''_i + f_i(M_i)$. Then $m'_i - m''_i \in f(M_i)$ and thus $\varphi'_\alpha(m'_i) - \varphi'_\alpha(m''_i) = \varphi'_\alpha(m'_i - m''_i)$ lies in $\varphi'_\alpha f_i(M_i) = f_j \varphi_\alpha(M_i) \subset f_j(M_j)$. It follows that $\chi_\alpha(m'_i + f_i(M_i)) = \chi_\alpha(m''_i + f_i(M_i))$, and therefore χ_α is well defined.

Definition 1.7. The representation $\mathrm{coker}\, f = (N_i, \chi_\alpha)_{i \in Q_0, \alpha \in Q_1}$ is called the **cokernel** of f.

Remark 1.4. The projections $\mathrm{proj}_i : M'_i \twoheadrightarrow \mathrm{coker}\, f_i$ induce a surjective morphism of representations[3]:

$$(\mathrm{proj}_i)_{i \in Q_0} : M' \twoheadrightarrow \mathrm{coker}\, f.$$

In category theory, kernels and cokernels are defined using the following *universal properties*:

Remark 1.5. Let $M \xrightarrow{g} N$ be a morphism. Then a kernel of g is a morphism $L \xrightarrow{f} M$ such that $gf = 0$, and given any morphism $X \xrightarrow{v} M$ such that $gv = 0$, there is a unique morphism $X \xrightarrow{u} L$ such that $fu = v$. We say that v factors through f:

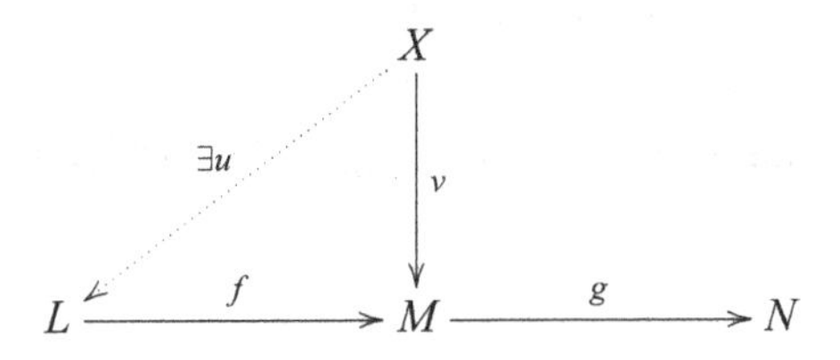

[2]The arrow $\hookrightarrow$ indicates that the morphism is injective.

[3]The arrow $\twoheadrightarrow$ indicates that the morphism is surjective.

Let us prove that our kernel from Definition 1.6 satisfies this universal property. So suppose that $g : M \to N$ is a morphism of representations of a quiver Q, let $L = \ker g$ as in Definition 1.6 and let f be the inclusion map $f : L \hookrightarrow M$. Then for every vertex $i \in Q_0$ and every $m_i \in L_i$, we have $g_i f_i(m_i) = g_i(m_i) = 0$, which shows that $gf = 0$.

Now suppose that $v : X \to M$ is a morphism of representations such that $gv = 0$. Let us use the notation $M = (M_i, \varphi_\alpha), L = (L_i, \psi_\alpha)$, and $X = (X_i, \chi_\alpha)$ for the three representations. Then for every $i \in Q_0$ and every $x_i \in X_i$, we have $v(x_i) \in \ker g_i = L_i$, so we can define a map $u : X \to L$ by $u_i(x_i) = v_i(x_i)$. It is clear that $fu = v$ so that the above diagram commutes, but we must check that u is actually a morphism of representations. So let $i \xrightarrow{\alpha} j$ be an arrow in Q, and let $x_i \in X_i$. Then, using the definitions of u and L and the fact that v is a morphism of representations, we get

$$\psi_\alpha u_i(x_i) = \varphi_\alpha v_i(x_i) = v_j \chi_\alpha(x_i) = u_j \chi_\alpha(x_i),$$

which shows that u is a morphism of representations. The fact that u is unique follows directly from the fact that f is the inclusion morphism.

Remark 1.6. Let $L \xrightarrow{f} M$ be a morphism. Then a cokernel of f is a morphism $M \xrightarrow{g} N$ such that $gf = 0$, and given any morphism $M \xrightarrow{v} X$ such that $vf = 0$, there is a unique morphism $N \xrightarrow{u} X$ such that $ug = v$. We say that v factors through g:

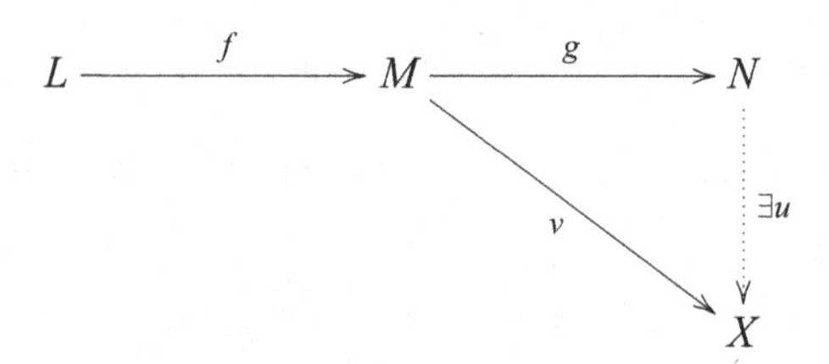

We leave it to the reader to prove that the two definitions of the cokernel agree.

Definition 1.8. A representation L is called a *subrepresentation* of a representation M if there is an injective morphism $i : L \hookrightarrow M$. In this situation, the *quotient representation* M/L is defined to be the cokernel of i.

Theorem 1.7 (First Isomorphism Theorem). *If $f : M \to N$ is a morphism of representations, then*

$$\operatorname{im} f \cong M/\ker f.$$

Proof. Let $M = (M_i, \varphi_\alpha)$. Then $\operatorname{im} f$ is the representation $\operatorname{im} f = (f(M_i), \psi_\alpha)$ whose maps are defined by $\psi_\alpha(f_i(m_i)) = f_j \varphi_\alpha(m_i)$ for every arrow $i \xrightarrow{\alpha} j$ in Q.

On the other hand, $M/\ker f$ is the representation $M/\ker f = (M_i/\ker f_i, \chi_\alpha)$ where $\chi_\alpha(m_i + \ker f_i) = \varphi_\alpha(m_i) + \ker f_j$.

Since each f_i is a linear map, it induces an isomorphism of vector spaces:

$$\bar{f_i} : M_i/\ker f_i \to f_i(M_i),\ m_i + \ker f_i \mapsto f_i(m_i).$$

Moreover, for every arrow $i \xrightarrow{\alpha} j$, we have $\psi_\alpha \bar{f_i} = \bar{f_j}\varphi_\alpha$, which shows that $\bar{f}$ is a morphism of representations. This completes the proof. □

Categories 3 *With the above definition of kernel and cokernel, we have that rep Q is an* abelian k-category. *This means that*

1. *rep Q is a k-*category, *that is, Hom(M,N) is a k-vector space for all $M,N \in$ rep Q, and the composition of morphisms is bilinear,*
2. *rep Q is* additive, *that is, rep Q has direct sums, there is a zero object $0 \in$ rep Q such that the identity morphism $1_0 \in$ Hom$(0,0)$ is the zero of the vector space Hom$(0,0)$, and*
3. *each morphism $f : M \to N$ in rep Q has a kernel $i : K \to M$ and a cokernel $p : N \to C$ such that the cokernel of i is isomorphic to the kernel of p.*

Observe that the condition coker $i \cong$ $\ker p$ *in 3 follows from the first isomorphism theorem.*

Next we introduce the notion of exact sequences which will be fundamental for the rest of the book.

Definition 1.9. A sequence of morphisms $L \xrightarrow{f} M \xrightarrow{g} N$ is called *exact at M* if $\operatorname{im} f = \ker g$. A sequence of morphisms

$$\cdots \longrightarrow M_1 \xrightarrow{f_1} M_2 \xrightarrow{f_2} M_3 \xrightarrow{f_3} \cdots$$

is called *exact* if it is exact at every M_i.

Definition 1.10. A **short exact sequence** is an exact sequence of the form

$$0 \longrightarrow L \xrightarrow{f} M \xrightarrow{g} N \longrightarrow 0\,.$$

Note that the sequence in Definition 1.10 is short exact if and only if f is injective, $\operatorname{im} f = \ker g$, and g is surjective.

Example 1.9. Let $f : M \to N$ be a morphism in rep Q. Then the sequence

$$0 \longrightarrow \ker f \xrightarrow{u} M \xrightarrow{f} N \xrightarrow{p} \operatorname{coker} f \longrightarrow 0\,,$$

where u is the inclusion of Remark 1.3 and p, the projection of Remark 1.4, is exact; and the sequence

$$0 \longrightarrow \ker f \xrightarrow{u} M \xrightarrow{q} M/\ker f \longrightarrow 0$$

is short exact.

Example 1.10. Let Q be the quiver $1 \longrightarrow 2$, and consider the three representations:

$$\begin{array}{ll} S(2) & (0 \longrightarrow k), \\ M & (k \xrightarrow{1} k), \\ S(1) & (k \longrightarrow 0). \end{array}$$

Then

$$0 \longrightarrow S(2) \xrightarrow{f} M \xrightarrow{g} S(1) \longrightarrow 0$$
$$0 \longrightarrow S(2) \xrightarrow{f'} S(1) \oplus S(2) \xrightarrow{g'} S(1) \longrightarrow 0,$$

where $f = (f_1, f_2) = (0, 1)$, $g = (g_1, g_2) = (1, 0)$, and $f' = (f_1', f_2') = (0, 1)$, $g' = (g_1', g_2') = (1, 0)$ are short exact sequences.

Definition 1.11. A morphism $f : L \to M$ is a called a **section** if there exists a morphism $h : M \to L$ such that $h \circ f = 1_L$.

A morphism $g : M \to N$ is a called a **retraction** if there exists a morphism $h : N \to M$ such that $g \circ h = 1_N$.

Definition 1.12. We say that a short exact sequence

$$0 \longrightarrow L \xrightarrow{f} M \xrightarrow{g} N \longrightarrow 0$$

splits if f is a section.

Example 1.11. In Example 1.10, the second short exact sequence splits, because the morphism $h' : S(1) \oplus S(2) \xrightarrow{(0,1)} S(2)$ verifies $h' \circ f' = 1_{S(2)}$. On the other hand, the first sequence does not split, since there is no nonzero morphism from M to $S(2)$; hence f cannot be a section.

Proposition 1.8. *Let*

$$0 \longrightarrow L \xrightarrow{f} M \xrightarrow{g} N \longrightarrow 0$$

be a short exact sequence in rep Q. Then

(a) *f is a section if and only if g is a retraction.*
(b) *If f is a section, then im $f(= \ker g)$ is a direct summand of M.*

Proof. First we show (a).

($\Rightarrow$) Suppose that f is a section. Then there exists $h \in \mathrm{Hom}(M, L)$ such that $h \circ f = 1_L$.

Define $h' : N \to M$ as follows: Let $n \in N$. Since g is surjective, there exists $m \in M$ such that $g(m) = n$. Choose one such m and define $h'(n) = m - f \circ h(m)$.

Since there may be different m to choose from, we must show that the definition of h' does not depend on the choice of m. Suppose that $m, m' \in M$ are such that $g(m) = g(m') = n$. We must show that $m - f \circ h(m) = m' - f \circ h(m')$. We have $g(m - m') = g(m) - g(m') = 0$, which shows that $m - m' \in \ker g$, and thus $m - m' \in \mathrm{im}\, f$. Therefore, there exists an $\ell \in L$ such that $f(\ell) = m - m'$, and consequently,

$$\begin{aligned} m - fh(m) - (m' - fh(m')) &= m - m' - fh(m - m') \\ &= m - m' - fhf(\ell) \\ &\overset{(*)}{=} m - m' - f(\ell) \\ &= 0, \end{aligned}$$

where the equation $(*)$ holds, because $h \circ f = 1_L$. This shows that h' is well defined.

Next we show that h' is a morphism. To do so, we need some notation. Let $L = (L_i, \varphi_\alpha)$, $M = (M_i, \varphi'_\alpha)$, and $N = (N_i, \varphi''_\alpha)$. Let $i \xrightarrow{\alpha} j$ be an arrow in Q_1. Then we have the following diagram:

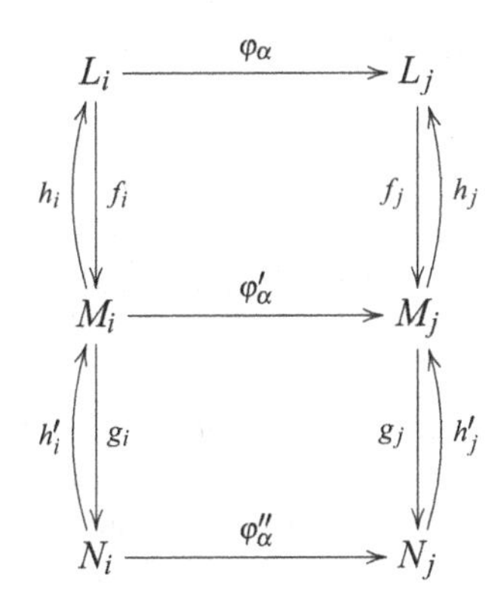

which is commutative with respect to the morphisms f, g and h, and we must show its commutativity with respect to h'. Let $n_i \in N_i$, and let $m_i \in M_i$ such that $g_i(m_i) = n_i$ as in the definition of h'. Then

$$\varphi'_\alpha h'_i(n_i) = \varphi'_\alpha(m_i - f_i h_i(m_i)) = \varphi'_\alpha(m_i) - \varphi'_\alpha f_i h_i(m_i),$$

which is shown to be equal to $\varphi'_\alpha(m_i) - f_j h_j \varphi'_\alpha(m_i)$ by using the commutativity of the diagram first for f and then for h.

On the other hand,

$$h'_j \varphi''_\alpha(n_i) = h'_j \varphi''_\alpha g_i(m_i) = h'_j g_j \varphi'_\alpha(m_i)$$

which is also equal to $\varphi'_\alpha(m_i) - f_j h_j \varphi'_\alpha(m_i)$, by definition of h'. This shows that $\varphi'_\alpha h'_i = h'_j \varphi''_\alpha$ and that h' is a morphism in rep Q.

Finally, we show that $gh' = 1_N$ and, consequently, that g is a retraction. To do so, let $n \in N$ and $m \in M$ such that $g(m) = n$. Then

$$gh'(n) = g(m - fh(m)) = g(m) - gf(h(m)) = g(m),$$

where the last equation holds, since $gf = 0$, because the sequence in the proposition is exact. But $g(m) = n$, and thus $gh' = 1_N$.

($\Leftarrow$) Suppose that g is a retraction. Then there is $h' \in \text{Hom}(N, M)$ such that $g \circ h' = 1_N$. Define $h : M \to L$ as follows. Let $m \in M$, and then $m - h'(g(m)) \in \ker g = \text{im}\, f$. Therefore there is $\ell \in L$ such that $f(\ell) = m - h'g(m)$, and this ℓ is unique, since f is injective. Define $h(m) = \ell$.

Clearly, $h \circ f = 1_L$.

In order to finish the proof, let us check that h is a morphism in rep Q. We will use the same notation as in the first part of the proof. Let $m_i \in M_i$ and let $\ell_i \in L_i$ such that

$$f_i(\ell_i) = m_i - h'_i g_i(m_i) \tag{1.1}$$

as in the definition of h. Then $\varphi_\alpha h_i(m_i) = \varphi_\alpha(\ell_i)$. On the other hand, by definition of h, we have $h_j \varphi'_\alpha(m_i) = \ell_j$ for a unique $\ell_j \in L_j$ with the property that

$$f_j(\ell_j) = \varphi'_\alpha(m_i) - h'_j g_j \varphi'_\alpha(m_i). \tag{1.2}$$

We must show that $\varphi_\alpha h_i = h_j \varphi'_\alpha$, so it suffices to show that $\varphi_\alpha(\ell_i) = \ell_j$. Now

$$\varphi'_\alpha(m_i) = \varphi'_\alpha f_i(\ell_i) + \varphi'_\alpha h'_i g_i(m_i) = f_j \varphi_\alpha(\ell_i) + h'_j g_j \varphi'_\alpha(m_i),$$

where the first identity holds because of (1.1) and the last identity holds because f, h and g are morphisms. Using this last identity to replace the first term on the right hand side of (1.2), we get

$$f_j(\ell_j) = f_j\varphi_\alpha(\ell_i),$$

and since f_j is injective, this implies that $\varphi_\alpha(\ell_i) = \varphi'_\alpha(m_i)$. This proves (a).

In order to prove (b), let $h' \in \operatorname{Hom}(N, M)$ be such that $gh' = 1_N$ and let $m = (m_i)_{i\in Q_0} \in M$. Then $m_i = h'_i g_i(m_i) + (m_i - h'_i g_i(m_i))$ with $h'_i g_i(m_i) \in \operatorname{im} h'_i$ and $(m_i - h'_i g_i(m_i)) \in \ker g_i$. Moreover, $\operatorname{im} h'_i \cap \ker g_i = \{0\}$, since $g \circ h' = 1_N$. Thus for each of the vector spaces M_i, we have a direct sum decomposition $M_i = \operatorname{im} h'_i \oplus \ker g_i$.

We still have to check that the maps of the representation M are the maps of the direct sum $\operatorname{im} h' \oplus \ker g$, that is, we must show that for each arrow $i \xrightarrow{\alpha} j$ in Q_1 we have

$$\varphi'_\alpha = \begin{bmatrix} \varphi'_\alpha|_{\operatorname{im} h'_i} & 0 \\ 0 & \varphi'_\alpha|_{\ker g_i} \end{bmatrix}. \tag{1.3}$$

If $m_i \in \ker g_i$, then $0 = \varphi''_\alpha g_i(m_i) = g_j\varphi'_\alpha(m_i)$, because g is a morphism. Therefore $\varphi'_\alpha(m_i) \in \ker g_j$ and thus the upper right block of the matrix in (1.3) is zero. If $m_i \in \operatorname{im} h'_i$, then there exists $n_i \in N_i$ such that $h'_i(n_i) = m_i$, and therefore $\varphi'_\alpha(m_i) = \varphi'_\alpha h'_i(n_i) = h'_j\varphi''_\alpha(n_i)$ is an element of $\operatorname{im}(h'_j)$. This shows that the lower left block of the matrix in (1.3) is zero, and therefore φ'_α is of the form (1.3), and we are done. □

Corollary 1.9. *If the sequence*

$$0 \longrightarrow L \xrightarrow{f} M \xrightarrow{g} N \longrightarrow 0$$

is split exact, then

$$M \cong L \oplus N.$$

Proof. Since f is injective, we have $L \cong f(L) \cong \ker g$, and, since g is surjective, the first isomorphism theorem implies $N \cong M/\ker g$. Now the result follows from Proposition 1.8. □

1.4 Hom Functors

We now want to introduce the Hom functors and study their effect on short exact sequences. First, let us recall the definition of functors.

Categories 4 *Let $\mathscr{C}, \mathscr{C}'$ be two k-categories. A* ***covariant functor*** $F : \mathscr{C} \to \mathscr{C}'$ *is a mapping that associates*

- *to each object $X \in \mathscr{C}$ an object $F(X) \in \mathscr{C}'$ and*
- *to each morphism $f : X \to Y$ in $\mathscr{C}$ a morphism $F(f): F(X) \to F(Y)$ in $\mathscr{C}'$,*

such that $F(1_X) = 1_{F(X)}$ and $F(g \circ f) = F(g) \circ F(f)$, for all objects X and all morphisms f and g in $\mathscr{C}$.
A **contravariant functor** $F : \mathscr{C} \to \mathscr{C}'$ is a mapping that associates:

- to each object $X \in \mathscr{C}$ an object $F(X) \in \mathscr{C}'$ and
- to each morphism $f : X \to Y$ in $\mathscr{C}$ a morphism $F(f): F(Y) \to F(X)$ in $\mathscr{C}'$,

such that $F(1_X) = 1_{F(X)}$ and $F(g \circ f) = F(f) \circ F(g)$, for all objects X and all morphisms f and g in $\mathscr{C}$.

Two very important functors are the Hom functors $\text{Hom}(X, -)$ and $\text{Hom}(-, X)$, where X is an arbitrary fixed object in the category $\mathscr{C}$. They are defined as follows:
$\text{Hom}(X, -)$ is the covariant functor from the category $\mathscr{C}$ to the category of k-vector spaces, which sends an object Y in $\mathscr{C}$ to the vector space $\text{Hom}(X, Y)$ of all morphisms from X to Y and which sends a morphism $(f : Y \to Z)$ in $\mathscr{C}$ to the map $f_* : \text{Hom}(X, Y) \longrightarrow \text{Hom}(X, Z)$, $f_*(g) = f \circ g$:

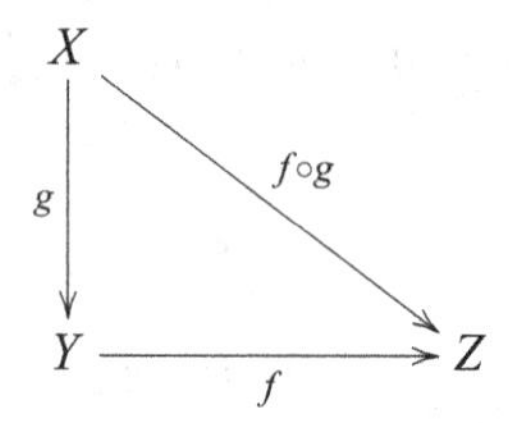

The map f_* is called the *push forward* of f.
$\text{Hom}(-, X)$ is the contravariant functor from the category $\mathscr{C}$ to the category of k-vector spaces, which sends an object Y in $\mathscr{C}$ to the vector space $\text{Hom}(Y, X)$ of all morphisms from Y to X and which sends a morphism $(f : Y \to Z)$ in $\mathscr{C}$ to the map $f_* : \text{Hom}(Z, X) \longrightarrow \text{Hom}(Y, X)$, $f^*(g) = g \circ f$:

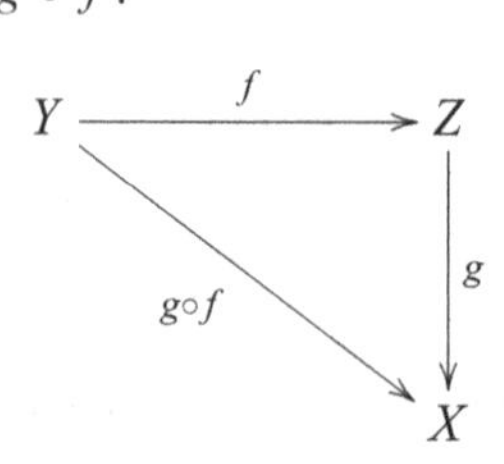

The map f^* is called the *pull back* of f.

Let us go back to our category rep Q of representations of the quiver Q. It turns out that applying the Hom functors to short exact sequences of representations yields new exact sequences of vector spaces.

Theorem 1.10. *Let Q be a quiver and $0 \longrightarrow L \xrightarrow{f} M \xrightarrow{g} N$ a sequence in rep Q. Then this sequence is exact if and only if for every representation $X \in$ rep Q, the sequence*

$$0 \longrightarrow \text{Hom}(X,L) \xrightarrow{f_*} \text{Hom}(X,M) \xrightarrow{g_*} \text{Hom}(X,N)$$

is exact.

Proof.

($\Rightarrow$) First, we show that f_* is injective. Suppose that there is $u \in \text{Hom}(X, L)$ such that $0 = f_*(u) = f \circ u$. Since f is injective, we can conclude that $u = 0$, and thus f_* is injective.

Next, we show that im $f_* = \ker g_*$. Let $u \in \text{Hom}(X, L)$. Then $g_* f_*(u) = g \circ f \circ u$, which is zero because $g \circ f = 0$. Hence $g_* f_* = 0$, and thus im $f_* \subset \ker g_*$.
On the other hand, let $v \in \text{Hom}(X, M)$ such that $v \in \ker g_*$. Then $0 = g_*(v) = g \circ v$. Using the universal property of the kernel of g (Remark 1.5) and the exactness of the sequence $0 \longrightarrow L \xrightarrow{f} M \xrightarrow{g} N$, this implies that v factors through f; thus there exists $u \in \text{Hom}(X, L)$ such that $v = f \circ u = f_*(u)$. Thus $v \in$ im f_*, and we have $\ker g_* \subset$ im f_*. Together with the other inclusion above, this implies $\ker g_* =$ im f_*.

($\Leftarrow$) First, we show that f is injective. Take $X = \ker f$, and let $i : X \hookrightarrow L$ be the inclusion morphism. Then $0 = f \circ i = f_*(i)$, and, since f_* is injective, this implies that $i = 0$. But since i is injective, it follows that $X = 0$, and thus f is injective.

Next, we show that im $f = \ker g$. Take $X = L$. Then $0 = g_* f_*(1_L) = g \circ f \circ 1_L = g \circ f$, and thus im $f \subset \ker g$.
On the other hand, take $X = \ker g$ and $i : X \hookrightarrow M$ the inclusion morphism. Then $0 = g \circ i = g_*(i)$ implies that $i \in \ker g_* =$ im f_*, and therefore there exists $u \in \text{Hom}(X, L)$ such that $i = f_*(u) = f \circ u$. Consequently $\ker g = i(X) \subset$ im f. Together with the other inclusion above, this implies $\ker g =$ im f, and we are done. □

Corollary 1.11. *A sequence*

$$0 \longrightarrow L \xrightarrow{f} M \xrightarrow{g} N \longrightarrow 0 \tag{1.4}$$

in rep Q is split exact if and only if for every $X \in$ rep Q, the sequence

$$0 \longrightarrow \operatorname{Hom}(X,L) \xrightarrow{f_*} \operatorname{Hom}(X,M) \xrightarrow{g_*} \operatorname{Hom}(X,N) \longrightarrow 0 \tag{1.5}$$

is exact.

Proof.

($\Rightarrow$) By Theorem 1.10, it suffices to show that g_* is surjective. Suppose that the sequence (1.4) is split exact. Then g is a retraction; hence there is $h \in \operatorname{Hom}(N, M)$ such that $gh = 1_N$.

Now for any $u \in \operatorname{Hom}(X, N)$, we have $hu \in \operatorname{Hom}(X, M)$ and $g_*(hu) = ghu = 1_N u = u$, which shows that g_* is surjective.

($\Leftarrow$) Suppose that for every $X \in$ rep Q, the sequence (1.5) is exact. Then it follows from Theorem 1.10 that the sequence

$$0 \longrightarrow L \xrightarrow{f} M \xrightarrow{g} N,$$

is exact. Taking $X = N$ and using the surjectivity of g_*, we see that there exists $h \in \operatorname{Hom}(N, M)$ such that

$$1_N = g_*(h) = gh,$$

which proves two facts:

1. g is surjective, which shows that the sequence (1.4) is exact, and
2. g is a retraction, which shows that the sequence (1.4) splits.

□

Remark 1.12. If

$$0 \longrightarrow L \xrightarrow{f} M \xrightarrow{g} N \longrightarrow 0$$

splits, then

$$0 \longrightarrow \operatorname{Hom}(X,L) \xrightarrow{f_*} \operatorname{Hom}(X,M) \xrightarrow{g_*} \operatorname{Hom}(X,N) \longrightarrow 0$$

splits too. Indeed, $gh = 1_N \Rightarrow g_* h_* = 1_{\operatorname{Hom}(X,N)}$.

There are *dual* versions of Theorem 1.10 and Corollary 1.11 involving the functor $\operatorname{Hom}(-, X)$. We state these results below, but leave the proofs as an exercise. Note that the order of the representations L, M, N is reversed in the Hom sequence, since $\operatorname{Hom}(-, X)$ is contravariant.

Theorem 1.13. *Let Q be a quiver and $L \xrightarrow{f} M \xrightarrow{g} N \longrightarrow 0$ a sequence in rep Q. Then this sequence is exact if and only if for every representation $X \in$ rep Q, the sequence*

$$0 \longrightarrow \mathrm{Hom}(N,X) \xrightarrow{g^*} \mathrm{Hom}(M,X) \xrightarrow{f^*} \mathrm{Hom}(L,X)$$

is exact.

Corollary 1.14. *A sequence*

$$0 \longrightarrow L \xrightarrow{f} M \xrightarrow{g} N \longrightarrow 0$$

in rep Q is split exact if and only if for every $X \in$ rep Q, the sequence

$$0 \longrightarrow \mathrm{Hom}(N,X) \xrightarrow{g^*} \mathrm{Hom}(M,X) \xrightarrow{f^*} \mathrm{Hom}(L,X) \longrightarrow 0$$

is exact.

Remark 1.15. If $0 \longrightarrow L \xrightarrow{f} M \xrightarrow{g} N \longrightarrow 0$ does not split, then f^* and g_* are not always surjective; see the example below. Nevertheless, one can extend the exact sequences of Theorems 1.10 and 1.13 to the right by introducing the *extension functors* $\mathrm{Ext}^i(X,-)$ and $\mathrm{Ext}^i(-,X)$; see Sect. 2.4.

Example 1.12. In Example 1.10 the short exact sequence

$$0 \longrightarrow S(2) \xrightarrow{f} M \xrightarrow{g} S(1) \longrightarrow 0$$

is non-split. Taking $X = S(1)$, and applying $\mathrm{Hom}(S(1),-)$, we get a morphism $g_*\colon \mathrm{Hom}(S(1),M) \to \mathrm{Hom}(S(1),S(1))$ which is not surjective since $\mathrm{Hom}(S(1),M) = 0$ and $\mathrm{Hom}(S(1),S(1)) \cong k$.

1.5 First Examples of Auslander–Reiten Quivers

We have already mentioned that the goal of representation theory of quivers is to study representations and morphisms in rep Q for a given quiver Q. To be even more ambitious, we may add the study of exact sequences in rep Q. In general, the so-called *Auslander–Reiten quiver* is a good first approximation of rep Q. In the case where the number of isoclasses of indecomposable representations is finite, the Auslander–Reiten quiver even provides complete information about rep Q.

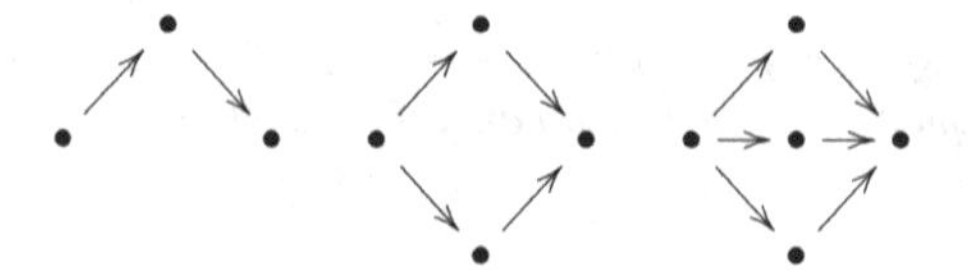

Fig. 1.1 Three different types of meshes

In this section, we give a sneak preview of Auslander–Reiten quivers. More examples will follow in Chap. 3, and for a more rigorous treatment see Chap. 7.

Let Q be a quiver. The Auslander–Reiten quiver of Q is a new quiver Γ_Q whose vertices are the isoclasses of indecomposable representations and whose arrows are given the by so-called *irreducible* morphisms. Roughly speaking, an irreducible morphism between two indecomposable representations is a morphism that does not factor nontrivially through another representation.

Recall that we can build any representation out of indecomposable ones; thus the vertices of the Auslander–Reiten quiver represent the building blocks for the representations.

The arrows of the Auslander–Reiten quiver, the irreducible morphisms, can be thought of the building blocks for morphisms in the sense that many (but in general not all!) morphisms are compositions of irreducible morphisms.

We also want to study short exact sequences of representations. As with morphisms, many of them (but in general not all!) are obtained by gluing together the so-called *almost split sequences*[4]. These almost split sequences are represented in the Auslander–Reiten quiver as meshes; see Fig. 1.1.

Example 1.13. Let Q be the quiver $1 \longrightarrow 2$. It follows from Exercise 1.4 that there are precisely three indecomposable representations (up to isomorphism), namely

$$\begin{array}{ccc} S(2) & M & S(1) \\ 0 \longrightarrow k & k \xrightarrow{\;1\;} k & k \longrightarrow 0. \end{array}$$

We have seen in Example 1.3 that

$$\begin{array}{c}\operatorname{Hom}(S(1),M)=0,\ \operatorname{Hom}(M,S(2))=0,\ \operatorname{Hom}(S(2),M)\cong k\\ \operatorname{Hom}(S(1),S(2))=0, \operatorname{Hom}(M,S(1))\cong k, \operatorname{Hom}(S(2),S(1))=0,\end{array}$$

and we conclude that there is only one non-split short exact sequence with indecomposable representations at the endpoints:

[4]Maurice Auslander and Idun Reiten introduced the concept of almost split sequences in [10].

$$0 \longrightarrow S(2) \longrightarrow M \longrightarrow S(1) \longrightarrow 0.$$

This sequence is actually an almost split sequence. Thus the Auslander–Reiten quiver consists of three vertices, two arrows, and one mesh and is of the form

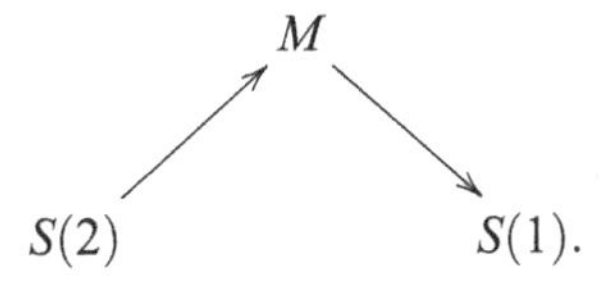

Remark 1.1. It is often convenient to have a shorthand notation for the representations, encoding the dimension vector and the maps. We will use the following notation throughout the whole book.

Let $Q_0 = \{1, 2, \ldots, n\}$ be the set of vertices of the quiver, let $M = (M_i, \varphi_\alpha)$ be an indecomposable representation of Q, and let $\underline{\dim}\, M = (d_{\Gamma}, d_2, \ldots, d_n)$ be its dimension vector. We describe the representation M as a configuration of digits using $1, 2, \ldots, n$ in such a way that the digit i appears exactly d_i times. Moreover, we arrange the digits in such a way that if there is an arrow $\alpha : i \to j$ such that the corresponding map $\varphi_\alpha : M_i \to M_j$ is nonzero, then the digit i is placed above the digit j. This notation has its limitations, but it is particularly useful if the isomorphism class of the representation M is determined by its dimension vector.

In the example above, we can picture the representation M by ${}^{2}_{1}$, meaning that $M_1 = k$ and $M_2 = k$ and the arrow is going downward from 1 to 2 and carries the identity map. In this notation, the whole Auslander–Reiten quiver would be

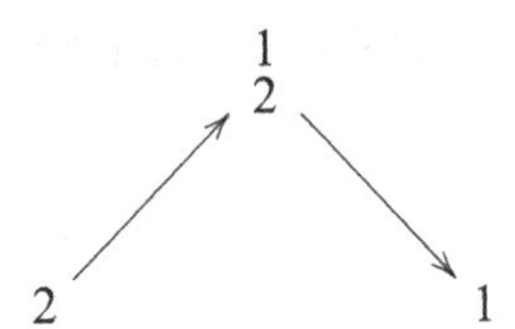

Example 1.14. Let Q be the quiver $1 \to 2 \leftarrow 3$. In this case, there are precisely six isoclasses of indecomposable representations, namely

$S(2)$	$P(1)$	$P(3)$
$0 \longrightarrow k \longleftarrow 0$	$k \xrightarrow{1} k \longleftarrow 0$	$0 \longrightarrow k \xleftarrow{1} k$
$I(2)$	$S(1)$	$S(3)$
$k \xrightarrow{1} k \xleftarrow{1} k$	$k \longrightarrow 0 \longleftarrow 0$	$0 \longrightarrow 0 \longleftarrow k$

or, using our symbolic notation,

$$S(2) = 2,\ P(1) = \begin{smallmatrix}1\\2\end{smallmatrix},\ P(3) = \begin{smallmatrix}3\\2\end{smallmatrix},\ I(2) = \begin{smallmatrix}1\ 3\\2\end{smallmatrix},\ S(1) = 1,\ S(3) = 3.$$

In this example, there are three almost split sequences:

$$0 \longrightarrow 2 \longrightarrow \begin{smallmatrix}1\\2\end{smallmatrix} \oplus \begin{smallmatrix}3\\2\end{smallmatrix} \longrightarrow \begin{smallmatrix}1\ 3\\2\end{smallmatrix} \longrightarrow 0$$

$$0 \longrightarrow \begin{smallmatrix}1\\2\end{smallmatrix} \longrightarrow \begin{smallmatrix}1\ 3\\2\end{smallmatrix} \longrightarrow 3 \longrightarrow 0$$

$$0 \longrightarrow \begin{smallmatrix}3\\2\end{smallmatrix} \longrightarrow \begin{smallmatrix}1\ 3\\2\end{smallmatrix} \longrightarrow 1 \longrightarrow 0$$

and the Auslander–Reiten quiver is of the form

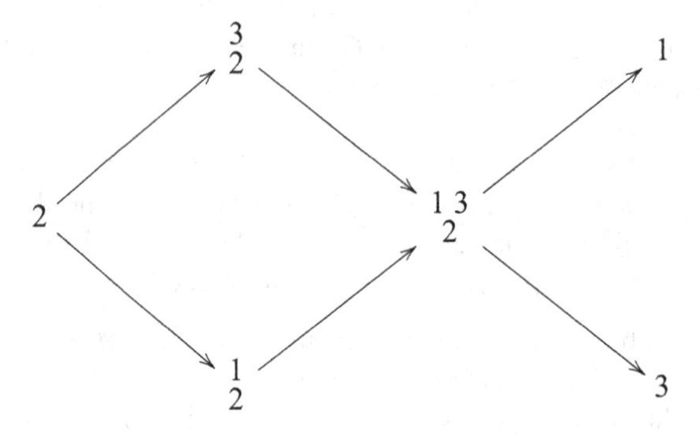

Let us point out that there are two further non-split short exact sequences with indecomposable end terms:

$$0 \longrightarrow 2 \longrightarrow \begin{smallmatrix}3\\2\end{smallmatrix} \longrightarrow 3 \longrightarrow 0$$

$$0 \longrightarrow 2 \longrightarrow \begin{smallmatrix}1\\2\end{smallmatrix} \longrightarrow 1 \longrightarrow 0$$

each of which can be obtained by “gluing the meshes” of two almost split sequences in the Auslander–Reiten quiver.

Problems

Exercises for Chap. 1

1.1. Let $M, M' \in \operatorname{rep} Q$. Show that the set of morphism $\operatorname{Hom}(M, M')$ is a k-vector space.

1.2. Let Q be the quiver $1 \longrightarrow 2 \longleftarrow 3$, and consider the representations

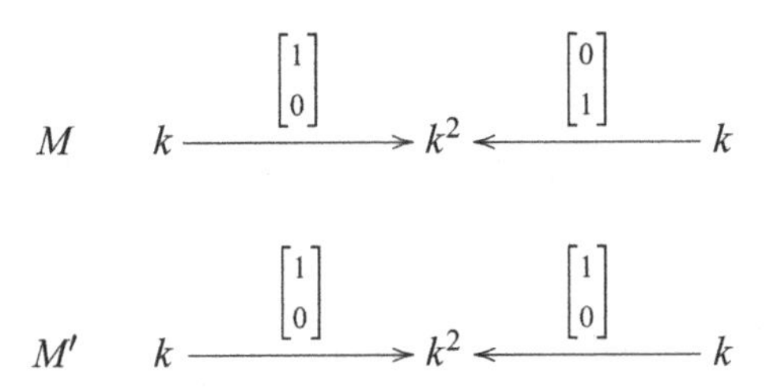

1. Show that M and M' are not indecomposable.
2. Show that M and M' are not isomorphic.

1.3. Let Q be the quiver $1 \longrightarrow 2 \longleftarrow 3$, and let M be the representation

$$k^2 \xrightarrow{\begin{bmatrix} 1 & 0 \\ 0 & 1 \\ 0 & 0 \end{bmatrix}} k^3 \xleftarrow{\begin{bmatrix} 1 \\ 0 \\ 0 \end{bmatrix}} k .$$

1. Write M as a direct sum of the indecomposable representations listed in Example 1.14.
2. Show that there is a non-split short exact sequence

$$0 \longrightarrow X \longrightarrow Y \longrightarrow Z \longrightarrow 0$$

such that $X \oplus Z = M$.

1.4. Find all indecomposable representations up to isomorphism of the quiver $1 \longrightarrow 2$. [Hint: Use the following theorem from Linear Algebra:]

Theorem 1.16. *Let $\phi : V_1 \to V_2$ be a linear map between finite-dimensional vector spaces and fix some bases for V_1 and V_2. Let r be the rank of ϕ. Then there exist isomorphisms of vector spaces $f_i : V_i \to V_i$ such that the matrix of $f_2 \circ \phi \circ f_1^{-1}$ with respect to the fixed bases is a diagonal matrix whose upper left $r \times r$ block is the identity matrix and all other entries are zero.*

1.5. Let Q be the quiver

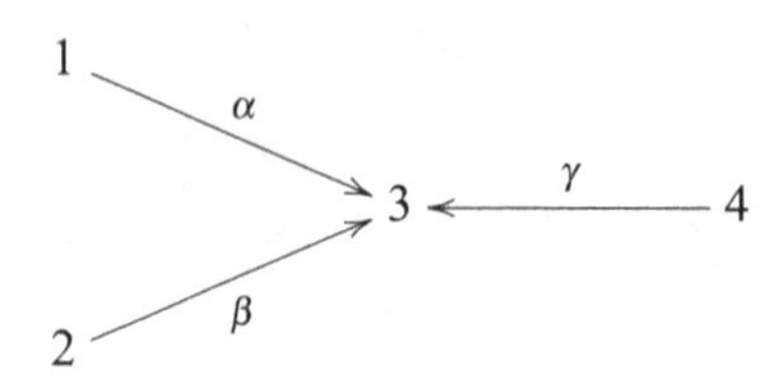

and consider the following representations:

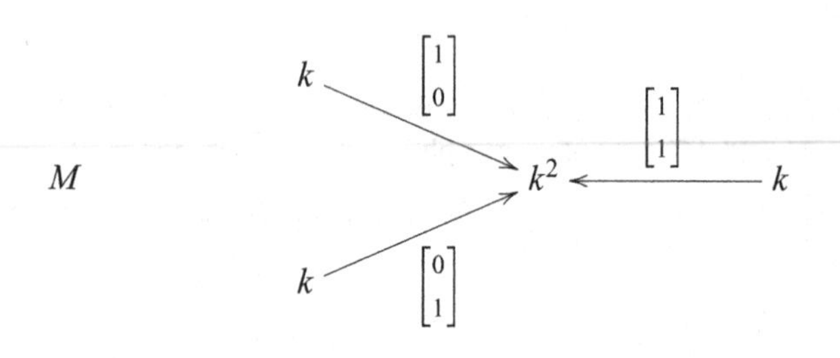

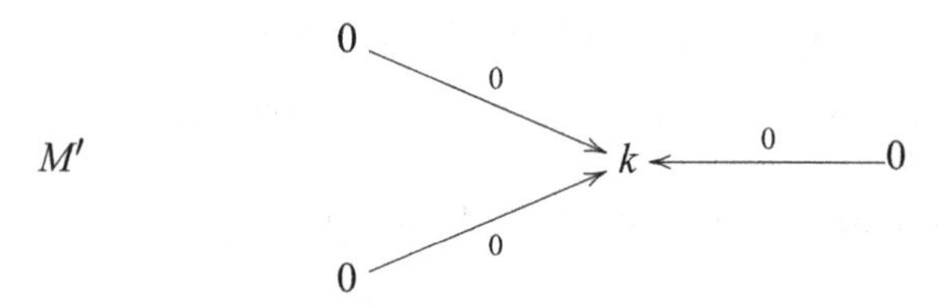

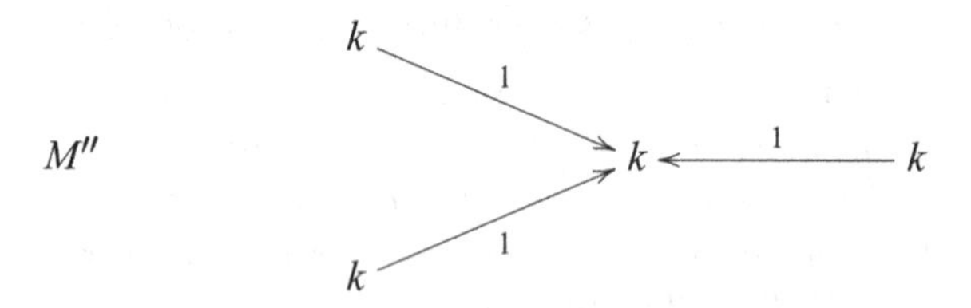

1. Show that

$$\begin{array}{ll} \mathrm{Hom}(M,M') = 0 & \mathrm{Hom}(M',M) \cong k^2 \\ \mathrm{Hom}(M,M'') \cong k^2 & \mathrm{Hom}(M'',M) = 0. \end{array}$$

2. Show that M is not isomorphic to $M' \oplus M''$.
3. Show that there is a short exact sequence:

$$0 \longrightarrow M' \longrightarrow M \longrightarrow M'' \longrightarrow 0.$$

1.6. Let Q be the quiver

$$1 \underset{\beta}{\overset{\alpha}{\leftleftarrows}} 2 .$$

For any $\lambda \in k \cup \{\infty\}$, define M_λ to be the representation:

$$k \underset{\lambda}{\overset{1}{\leftleftarrows}} k \text{ ; if } \lambda \in k;$$

$$k \underset{1}{\overset{0}{\leftleftarrows}} k \text{ ; if } \lambda = \infty.$$

1. Show that each M_λ is indecomposable.
2. Show that $M_\lambda \cong M_\mu$ if and only if $\lambda = \mu$. In particular, the number of indecomposable representations depends on the choice of the field k.
3. Show that $\mathrm{Hom}(M_\lambda, M_\mu) = 0$ if $\lambda \neq \mu$.
4. Show that for each λ there is a short exact sequence

$$0 \longrightarrow 1 \longrightarrow M_\lambda \longrightarrow 2 \longrightarrow 0,$$

where 1 and 2 are the representations

$$1 : k \underset{0}{\overset{0}{\leftleftarrows}} 0 \quad , \quad 2 : 0 \underset{0}{\overset{0}{\leftleftarrows}} k .$$

1.7. Let $f : L \to M$ be a morphism of representations.

1. Show that the cokernel of f together with the projection $\pi : M \to \operatorname{coker} f$ satisfies the universal property of Remark 1.6.
2. Show that if $g : M \to N$ is a morphism satisfying the universal property of Remark 1.6, then $N \cong \operatorname{coker} f$.

1.8. Let M, M', N be representations of Q and let $f : M \to N, g : M' \to N$ be morphisms. Define the *fiber product* (or *pull back*) of f and g as

$$X = \{(a, b) \mid a \in M, b \in M', \text{ such that } f(a) = g(b)\},$$

and define the projections $\pi_1(a, b) = a$ and $\pi_2(a, b) = b$.

1. Show that X is a subrepresentation of $M \oplus M'$ and that there is a commutative diagram:

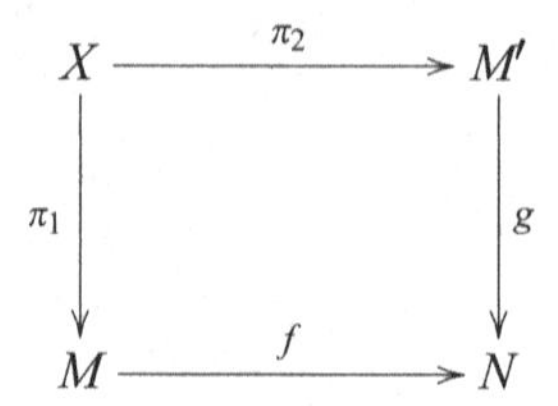

2. Show that if f is injective then π_2 is injective.
3. Show that if f is surjective then π_2 is surjective.
4. Now suppose $0 \longrightarrow L \xrightarrow{h} M \xrightarrow{f} N \longrightarrow 0$ is a short exact sequence and define $h' : L \to X$ by $h'(n) = (h(n), 0)$. Show that the following diagram is commutative with exact rows:

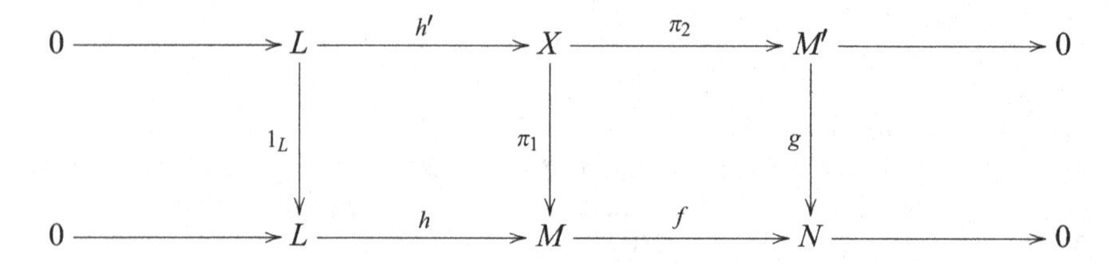

1.9. Let L, M, M' be representations of Q and let $f : L \to M, g : L \to M'$ be morphisms. Define the *amalgamated sum* (or *push out*) of f and g as

$$X = (M \oplus M')/\{(f(\ell), -g(\ell)) \mid \ell \in L\},$$

and define the morphisms $u_1 \colon M \to X$ and $u_2 \colon M' \to X$ by $u_1(m) = \overline{(m, 0)}$ and $u_2(m') = \overline{(0, m')}$, where $\overline{(a, a')}$ denotes the class of $(a, a') \in M \oplus M'$ in X.

1. Show that there is a commutative diagram:

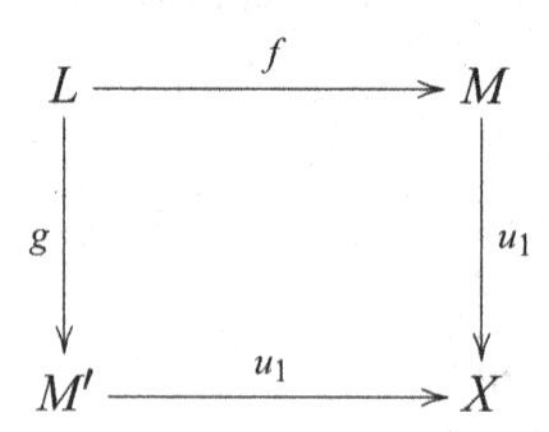

2. Now suppose $0 \longrightarrow L \xrightarrow{f} M \xrightarrow{h} N \longrightarrow 0$ is a short exact sequence and define $h' : X \to N$ by $h'\overline{(m, m')} = h(m)$. Show that the following diagram is commutative with exact rows:

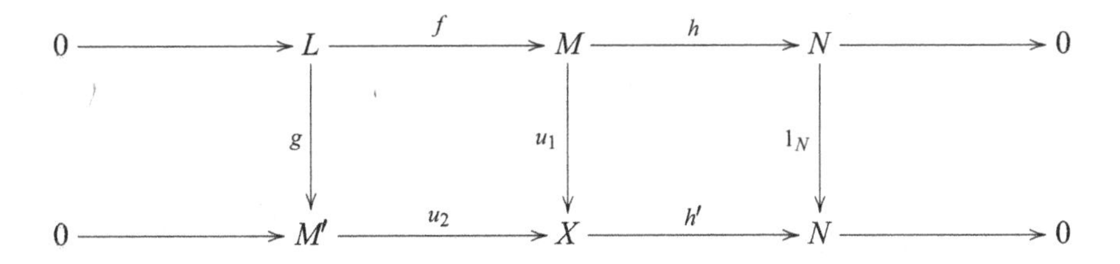

1.10. Write out the morphisms of the 5 short exact sequences in Example 1.14. Show that the almost split sequence

$$0 \longrightarrow 2 \xrightarrow{h'} \begin{smallmatrix}1\\2\end{smallmatrix} \oplus \begin{smallmatrix}3\\2\end{smallmatrix} \xrightarrow{\pi_2} \begin{smallmatrix}1\ 3\\2\end{smallmatrix} \longrightarrow 0$$

is obtained from the short exact sequence

$$0 \longrightarrow 2 \xrightarrow{h} \begin{smallmatrix}1\\2\end{smallmatrix} \xrightarrow{f} 1 \longrightarrow 0$$

as in Exercise 1.8 via the fiber product of f and the irreducible morphism $g : \begin{smallmatrix}1\ 3\\2\end{smallmatrix} \to 1$.

Chapter 2
Projective and Injective Representations

Projective representations and injective representations are key concepts in representation theory. A representation P is called *projective* if the functor $\operatorname{Hom}(P, -)$ maps surjective morphisms to surjective morphisms. Dually a representation I is called *injective* if the functor $\operatorname{Hom}(-, I)$ maps injective morphisms to injective morphisms. The terminology comes from the property that for any representation M there is a projective representation P_0 such that there exists a surjective morphism (a "projection"):

$$p_0 : P_0 \twoheadrightarrow M.$$

Dually, for any representation M, there is an injective representation I_0 such that there exists an injective morphism:

$$i_0 : M \hookrightarrow I_0.$$

If M is not projective itself, then the morphism p_0 above will have a kernel, and we can find another projective P_1 such that there exists a surjective morphism p_1 from P_1 to the kernel of p_0. Iterating this procedure yields an exact sequence

$$\cdots \longrightarrow P_3 \xrightarrow{p_3} P_2 \xrightarrow{p_2} P_1 \xrightarrow{p_1} P_0 \xrightarrow{p_0} M \longrightarrow 0$$

where each P_i is a projective representation. Such a sequence is called a *projective resolution*. We think of projective resolutions as a way to approximate the representation M by projective representations. Often it is possible to deduce properties of M from a projective resolution of M.

R. Schiffler, *Quiver Representations*, CMS Books in Mathematics,
DOI 10.1007/978-3-319-09204-1_2

Dually, we have *injective resolutions*, thus exact sequences of the form

$$0 \longrightarrow M \xrightarrow{i_0} I_0 \xrightarrow{i_1} I_1 \xrightarrow{i_2} I_2 \xrightarrow{i_3} I_3 \longrightarrow \cdots$$

where each I_i is an injective representation.

For representations of quivers without oriented cycles the situation is very simple. We will see that every quiver representation has a projective resolution of the form

$$0 \longrightarrow P_1 \longrightarrow P_0 \longrightarrow M \longrightarrow 0\,,$$

and an injective resolution of the form

$$0 \longrightarrow M \longrightarrow I_0 \longrightarrow I_1 \longrightarrow 0\,.$$

We will use this result to show that every subrepresentation of a projective representation is projective. Categories with this property are called *hereditary*.

Moreover, it is very easy to write down all indecomposable projective representations of a quiver Q without oriented cycles. There is exactly one indecomposable projective representation $P(i)$ for each vertex $i \in Q_0$, and this representation $P(i)$ is given by the paths in Q starting at the vertex i. Dually, there is exactly one indecomposable injective representation $I(i)$ for each vertex $i \in Q_0$, and $I(i)$ is given by the paths ending at the vertex i.

We will see later in Chap. 4 that each quiver defines an algebra A, the *path algebra of the quiver*, whose basis consist of the set of all paths in Q. We will also see that we can consider the algebra A as a representation of Q and that this representation is isomorphic to the direct sum of the indecomposable projective representations, thus $A \cong \oplus_{i \in Q_0} P(i)$ as representations of Q.

In the current chapter, we also introduce the Auslander–Reiten translation τ, which is crucial to Auslander–Reiten theory and Auslander–Reiten quivers. It is defined in a rather curious way by taking the beginning of a projective resolution of M

$$P_1 \xrightarrow{p_1} P_0 \longrightarrow M \longrightarrow 0$$

and then setting $\tau M = \ker \nu p_1$, where ν is the so-called Nakayama functor. This functor maps projective representations to injective representations, and therefore we obtain the beginning of an injective resolution:

$$0 \longrightarrow \tau M \longrightarrow \nu P_1 \xrightarrow{\nu p_1} \nu P_0.$$

Throughout this chapter, and the rest of the book, the very natural notion of paths in a quiver will be essential. Here is a formal definition.

Definition 2.1. Let $Q = (Q_0, Q_1, s, t)$ be a quiver, $i, j \in Q_0$. A **path** c from i to j of length ℓ in Q is a sequence

$$c = (i\,|\alpha_1, \alpha_2, \ldots, \alpha_\ell|\,j)$$

with $\alpha_h \in Q_1$ such that

$$\begin{aligned} s(\alpha_1) &= \quad i, \\ s(\alpha_h) &= t(\alpha_{h-1}), \text{ for } h = 2, 3, \ldots, \ell, \\ t(\alpha_\ell) &= \quad j. \end{aligned}$$

Thus a path from i to j is a way to go from vertex i to vertex j in the quiver Q, where we are only allowed to walk along an arrow in the direction to which it is pointing.

Example 2.1. In the quiver

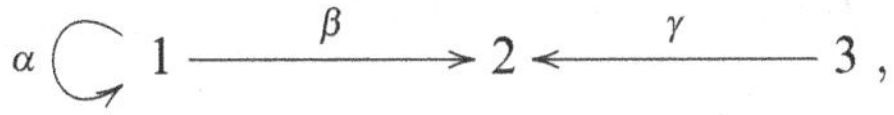

$$\alpha \circlearrowright 1 \xrightarrow{\beta} 2 \xleftarrow{\gamma} 3\,,$$

we have that $(1|\alpha|1)$, $(1|\alpha, \beta|2)$, $(1|\alpha, \alpha, \beta|2)$ are paths, but $(1|\alpha, \beta, \gamma|2)$ is not.

Example 2.2.

1. The **constant path** (or lazy path) $(i||i)$ at vertex i is the path of length $\ell = 0$ which never leaves the vertex i. We denote this path e_i.
2. An arrow $i \xrightarrow{\alpha} j$ is a path $(i|\alpha|j)$ of length one. If $i = j$ then

$$i \circlearrowright \alpha$$

is called a **loop**.
3. A path of the form

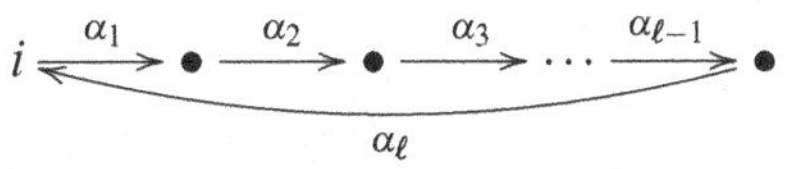

$$i \xrightarrow{\alpha_1} \bullet \xrightarrow{\alpha_2} \bullet \xrightarrow{\alpha_3} \cdots \xrightarrow{\alpha_{\ell-1}} \bullet \quad (\alpha_\ell : \bullet \to i)$$

given by $(i|\alpha_1, \alpha_2, \ldots, \alpha_\ell|i)$ is called an **oriented cycle**. Thus a loop is an oriented cycle of length one.

2.1 Simple, Projective, and Injective Representations

Let Q be a quiver without oriented cycles. For every vertex i in Q, we will now define three representations: the simple, the projective, and the injective representation at i.

We will show that in the category rep Q, these representations are respectively simple, projective, or injective objects in the categorical sense.

Definition 2.2. Let i be a vertex of Q. Define representations $S(i)$, $P(i)$, and $I(i)$ as follows:

(a) $S(i)$ is of dimension one at vertex i, and zero at every other vertex; thus

$$S(i) = (S(i)_j, \varphi_\alpha)_{j\in Q_0,\alpha\in Q_1}, \quad \text{where}$$

$$S(i)_j = \begin{cases} k \text{ if } i = j, \\ 0 \text{ otherwise,} \end{cases} \quad \text{and}$$

$$\varphi_\alpha = 0 \text{ for all arrows } \alpha.$$

$S(i)$ is called the **simple representation** at vertex i.

(b)

$$P(i) = (P(i)_j, \varphi_\alpha)_{j\in Q_0,\alpha\in Q_1}$$

where $P(i)_j$ is the k-vector space with basis the set of all paths from i to j in Q; so the elements of $P(i)_j$ are of the form $\sum_c \lambda_c c$, where c runs over all paths from i to j, and $\lambda_c \in k$;
and if $j \xrightarrow{\alpha} \ell$ is an arrow in Q, then $\varphi_\alpha : P(i)_j \to P(i)_\ell$ is the linear map defined on the basis by composing the paths from i to j with the arrow $j \xrightarrow{\alpha} \ell$.

More precisely, the arrow α induces an injective map between the bases

$$\begin{array}{ccc} \text{basis of } P(i)_j & \longrightarrow & \text{basis of } P(i)_\ell \\ c = (i|\beta_1, \beta_2, \ldots, \beta_s|j) & \longmapsto & c\,\alpha = (i|\beta_1, \beta_2, \ldots, \beta_s, \alpha|\ell) \end{array}$$

and φ_α is defined by

$$\varphi_\alpha\left(\sum_c \lambda_c\, c\right) = \sum_c \lambda_c\, c\,\alpha.$$

$P(i)$ is called the **projective representation** at vertex i.

(c)

$$I(i) = (I(i)_j, \varphi_\alpha)_{j\in Q_0,\alpha\in Q_1}$$

where $I(i)_j$ is the k-vector space with basis the set of all paths from j to i in Q; so the elements of $I(i)_j$ are of the form $\sum_c \lambda_c c$, where c runs over all paths from j to i, and $\lambda_c \in k$;
and if $j \xrightarrow{\alpha} \ell$ is an arrow in Q, then $\varphi_\alpha : I(i)_j \to I(i)_\ell$ is the linear map defined on the basis by deleting the arrow $j \xrightarrow{\alpha} \ell$ from those paths from j to i which start with α and sending to zero the paths that do not start with α.

More precisely, the arrow α induces a surjective map f between the bases

$$\begin{array}{ccc} \text{basis of } I(i)_j & \xrightarrow{f} & \text{basis of } I(i)_\ell \\ c = (j|\beta_1, \beta_2, \ldots, \beta_s|i) & \longmapsto & \begin{cases} (\ell|\beta_2, \ldots, \beta_s|i) \text{ if } \beta_1 = \alpha, \\ 0 \qquad\qquad \text{otherwise;} \end{cases} \end{array}$$

and φ_α is defined by

$$\varphi_\alpha\left(\sum_c \lambda_c c\right) = \sum_c \lambda_c f(c).$$

$I(i)$ is called the **injective representation** at vertex i.

Note that we need the hypothesis that Q has no oriented cycles, because, otherwise, there would be a vertex i such that $P(i)$ is infinite-dimensional, and hence not a representation in rep Q. For example, if Q is the quiver $1 \rightleftarrows 2$ then $P(1)$ and $P(2)$ would be infinite-dimensional.

The following remark will be very useful later on.

Remark 2.1. Let $P(i) = (P(i)_j, \varphi_\alpha)$ be the projective representation at vertex i and let c be a path starting at i, say

$$c = (i|\beta_1, \beta_2, \ldots, \beta_\ell|j).$$

Then we can define the map

$$\varphi_c \colon P(i)_i \longrightarrow P(i)_j \qquad \varphi_c = \varphi_{\beta_\ell} \cdots \varphi_{\beta_2}\varphi_{\beta_1}$$

as the composition of the maps in the representation $P(i)$ along the path c. Then, if e_i denotes the constant path at vertex i, it follows from the definition of $P(i)$ that

$$\varphi_c(e_i) = c. \tag{2.1}$$

Remark 2.2. Simple projectives and simple injectives:

(1) The projective representation at vertex i is the simple representation at vertex i if and only if there is no arrow α in Q such that $s(\alpha) = i$. Such vertices are called *sinks* of the quiver Q. Thus

$$S(i) = P(i) \iff i \text{ is a sink in } Q.$$

(2) The injective representation at vertex i is the simple representation at vertex i if and only if there is no arrow α in Q such that $t(\alpha) = i$. Such vertices are called *sources* of the quiver Q. Thus

$$S(i) = I(i) \iff i \text{ is a source in } Q.$$

In the following two examples, we use matrix notation to describe projective and injective representations. We use the isomorphism and not the equality symbol since there are many other possible descriptions for these representations; see Exercise 2.1.

Example 2.3. Let Q be the quiver

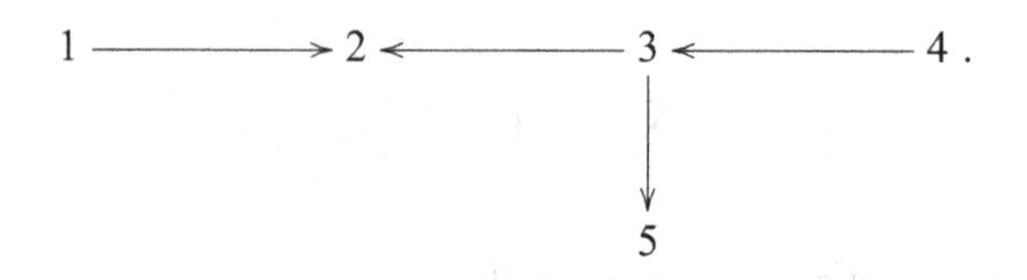

Then

$$\begin{array}{ccccccc}
S(3) \cong 0 & \longrightarrow & 0 & \longleftarrow & k & \longleftarrow & 0\,, \\
 & & & & \downarrow & & \\
 & & & & 0 & &
\end{array}$$

$$\begin{array}{ccccccc}
P(3) \cong 0 & \longrightarrow & k & \xleftarrow{\;1\;} & k & \longleftarrow & 0\,, \\
 & & & & \downarrow{\scriptstyle 1} & & \\
 & & & & k & &
\end{array}$$

$$\begin{array}{ccccccc}
I(3) \cong 0 & \longrightarrow & 0 & \longleftarrow & k & \xleftarrow{\;1\;} & k\,. \\
 & & & & \downarrow & & \\
 & & & & 0 & &
\end{array}$$

The quiver in the next example contains parallel paths. As a result the indecomposable projective modules can be of dimension greater than 1 at a single vertex.

Example 2.4. Let Q be the quiver

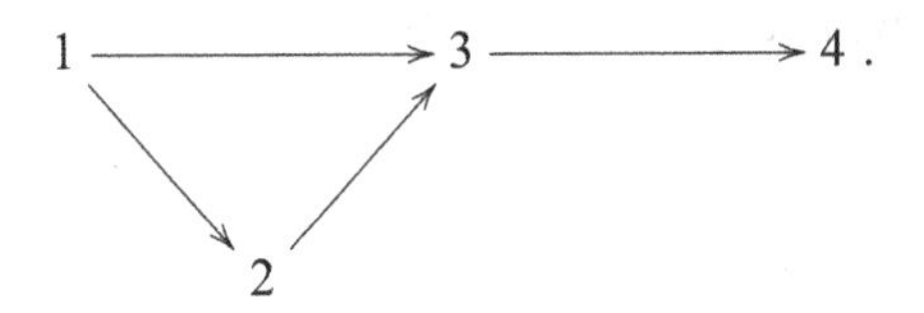

Then

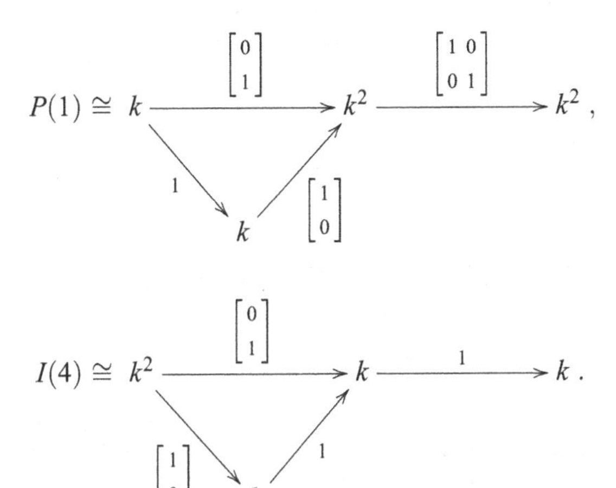

In category theory, a **projective** object is an object P such that the Hom functor $\mathrm{Hom}(P, -)$ maps surjective morphisms to surjective morphisms. The following proposition shows that the representations $P(i)$ satisfy this condition, which is the reason why we call them projective.

Proposition 2.3. *Let $g\colon M \to N$ be a surjective morphism between representations of Q, and let $P(i)$ be the projective representation at vertex i. Then the map*

$$g_*\colon Hom(P(i), M) \longrightarrow Hom(P(i), N)$$

is surjective.

In other words, if $f\colon P(i) \to N$ is any morphism, then there exists a morphism $h\colon P(i) \to M$ such that the diagram

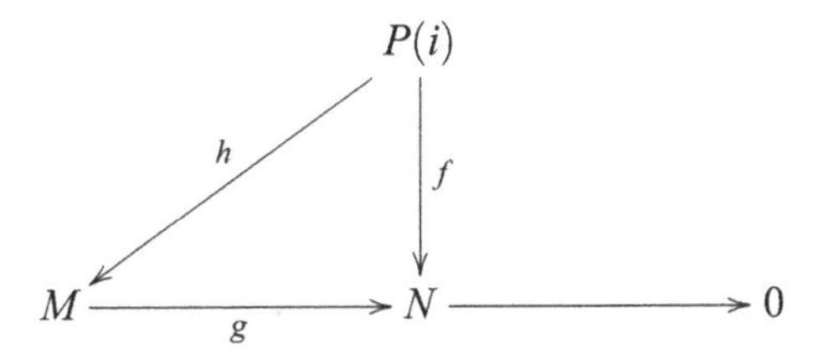

commutes, that is, $f = g \circ h = g_*(h)$.

Proof. Exercise 2.4. □

Corollary 2.4. *If P is projective, then any exact sequence of the form*

$$0 \longrightarrow L \longrightarrow M \xrightarrow{g} P \longrightarrow 0$$

splits.

Proof. Use Proposition 2.3 with the identity morphism $f = 1_P$ to get the commutative diagram:

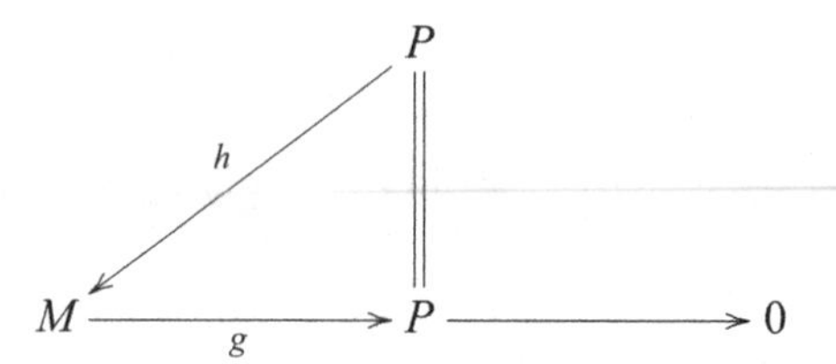

Therefore $1 = g \circ h$, and g is a retraction. □

A dual statement holds for the injective representations. In category theory, an **injective** object is an object I such that the Hom functor $\mathrm{Hom}(-, I)$ maps injective morphisms to surjective morphisms. The next proposition shows that the representations $I(i)$ satisfy this condition, which is the reason why we call them injective.

Proposition 2.5. *Let $g: L \to M$ be an injective morphism between representations of Q, and let $I(i)$ be the injective representation at vertex i. Then the map*

$$g^*: Hom(M, I(i)) \longrightarrow Hom(L, I(i))$$

is surjective.

In other words, if $f: L \to I(i)$ is any morphism, then there exists a morphism $h: M \to I(i)$ such that the diagram

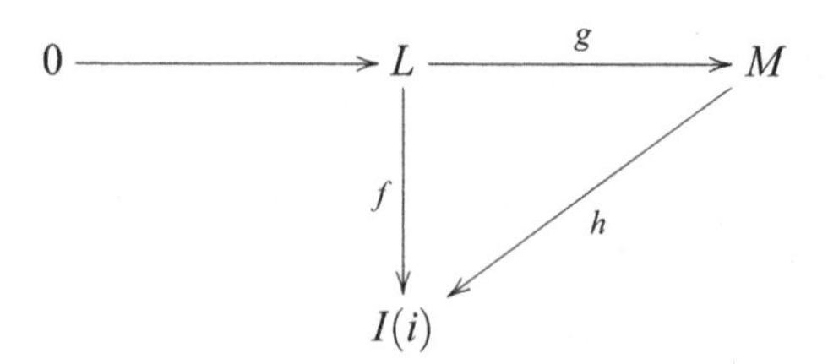

commutes, that is, $f = h \circ g = g^*(h)$.

Proof. Exercise 2.5. □

Corollary 2.6. *If I is injective then any exact sequence of the form*

$$0 \longrightarrow I \xrightarrow{g} M \longrightarrow N \longrightarrow 0$$

splits.

Proof. Use Proposition 2.5 with the identity morphism $f = 1_I$ to get a commutative diagram:

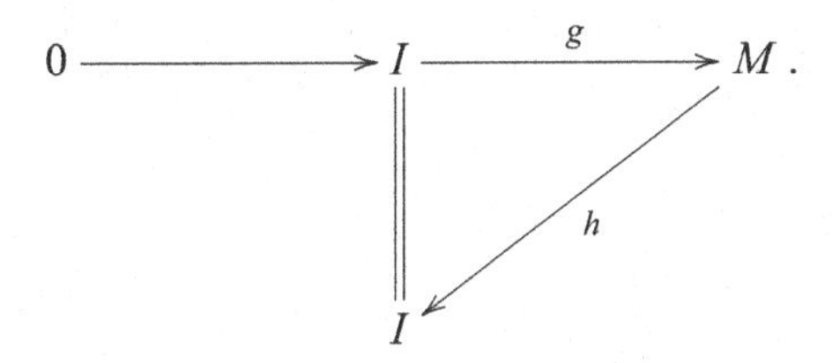

Thus $1_I = h \circ g$, and g is a section. □

Finally, a **simple** object in a category is a nonzero object S that has no proper subobjects. The representations $S(i)$ have this property, hence their name.

The next proposition states that sums of projective objects are projective and that summands of projective objects are projective. We state the result for the category rep Q, but the proof holds in any additive category.

Proposition 2.7. *(1) Let P and P' be representations of Q. Then*

$$P \oplus P' \text{ is projective} \iff P \text{ and } P' \text{ are projective.}$$

(2) Let I and I' be representations of Q. Then

$$I \oplus I' \text{ is injective} \iff I \text{ and } I' \text{ are injective.}$$

Proof. We only show (1) since the proof of (2) is similar.

($\Rightarrow$) Let $g : M \to N$ be surjective in rep Q and let $f : P \to N$ be any morphism in rep Q. Consider the following diagram:

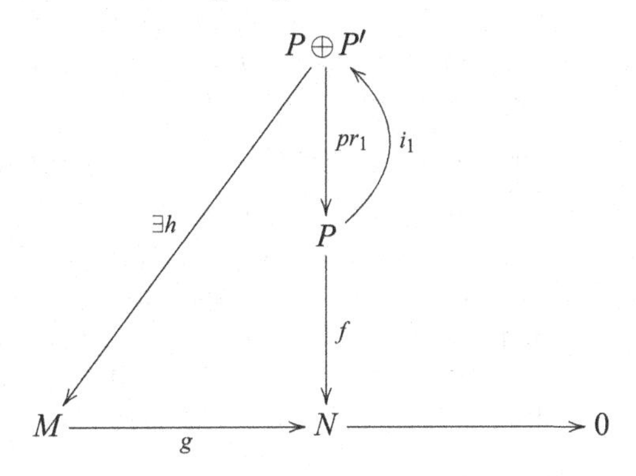

where pr_1 denotes the projection on the first summand and i_1 is the canonical injection. Clearly, $pr_1 \circ i_1 = 1_P$. Since $P \oplus P'$ is projective, there exists a map $h : P \oplus P' \to M$ such that $g\,h = f\,pr_1$. Therefore

$$g\,h\,i_1 = f\,pr_1\,i_1 = f\,1_P = f.$$

Now we can define $h' : P \to M$ as $h' = h\,i_1$, and we have $g\,h' = f$. This shows that P is projective. One can show in a similar way that P' is projective.

($\Leftarrow$) Let $g : M \to N$ be a surjective morphism in rep Q and let $f : P \oplus P' \to N$ be any morphism in rep Q. Consider the following commutative diagram:

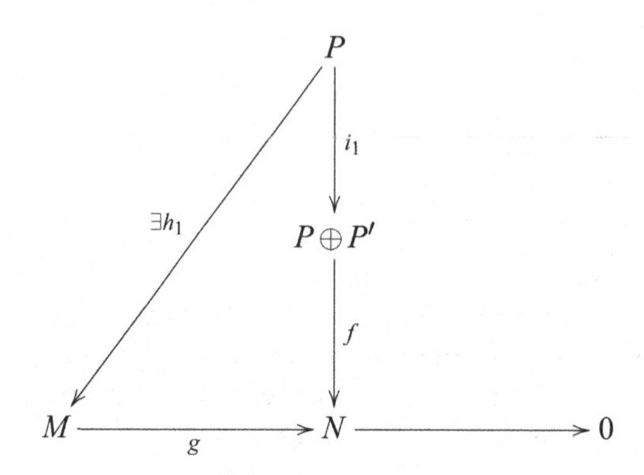

where i_1 denotes the canonical injection and, since P is projective, there exists a morphism h_1 such that $g\,h_1 = f\,i_1$. By symmetry, there also exists a morphism $h_2 : P' \to M$ such that $g\,h_2 = f\,i_2$. Define $h = (h_1, h_2) : P \oplus P' \to M$ by $h(p + p') = h_1(p) + h_2(p')$. Then

$$g\,h(p + p') = g\,h_1(p) + g\,h_2(p') = f\,i_1(p) + f\,i_2(p') = f(p + p'),$$

which shows that $P \oplus P'$ is projective. □

Proposition 2.7 implies that if we know the indecomposable projective, respectively injective, representations, then we know all projective, respectively injective, representations. The next step is to show that the representations $P(i)$ and $I(i)$ are in fact indecomposable. We will see later in Corollary 2.21 that there are no other indecomposable projective or injective representations.

Proposition 2.8. *The representations $S(i)$, $P(i)$, and $I(i)$ are indecomposable.*

Proof. For $S(i)$ this follows directly from the fact that $S(i)$ is simple. Let us prove the result for the projective representation $P(i) = (P(i)_j, \varphi_\alpha)_{j \in Q_0, \alpha \in Q_1}$. Since Q has no oriented cycles, we have $P(i)_i = k$. Suppose that $P(i) = M \oplus N$ for some $M, N \in \text{rep}\,Q$. Then we may suppose without loss of generality that $P(i)_i = M_i$ and $N_i = 0$. Let ℓ be a vertex of Q such that $N_\ell \neq 0$. Now, $P(i)_\ell$ has a basis consisting of the paths from i to ℓ in Q. Let $c = (i|\beta_1, \dots, \beta_s|\ell)$ be such a path.

Let $\varphi_c = \varphi_{\beta_s} \cdots \varphi_{\beta_1}$ denote the composition of the linear maps of the representation $P(i)$ along the path c. Then, since $P(i)$ is the direct sum of M and N, the map

$$\varphi_c : M_i \oplus 0 \longrightarrow M_\ell \oplus N_\ell$$

sends the unique basis element e_i of M_i to an element $\varphi_c(e_i)$ of M_ℓ. But from Remark 2.1 we know that $\varphi_c(e_i) = c$; thus every basis element c of $P(i)_\ell$ lies in M_ℓ, a contradiction.

The proof for $I(i)$ is similar. □

The following proposition shows that the simple representations $S(i)$ form a complete set of simple representations in rep Q, up to isomorphism.

Proposition 2.9. *A representation of Q is simple if and only if it is isomorphic to $S(i)$, for some $i \in Q_0$.*

Proof. It is clear that the $S(i)$ are simple representations. Conversely, let $M = (M_i, \varphi_\alpha)$ be any representation of Q. We want to show that there is a vertex i such that $S(i)$ is a subrepresentation of M, and we have to choose this vertex i carefully. We do not want to have a nonzero map in the representation M that starts at the vertex i. For example, if i is a sink in the quiver, we have what we want. But on the other hand, we also need the representation M to be nonzero at the vertex i. This leads us to pick i as follows.

Let $i \in Q_0$ such that $M_i \neq 0$ and $M_j = 0$, whenever there is an arrow $i \xrightarrow{\alpha} j$ in Q. Note that such a vertex exists since Q has no oriented cycles. Choose any injective linear map $f_i : S(i)_i \cong k \to M_i$, and extend it trivially to a morphism $f : S(i) \to M$ by letting $f_j = 0$ if $i \neq j$. Note that f actually is a morphism since the diagram

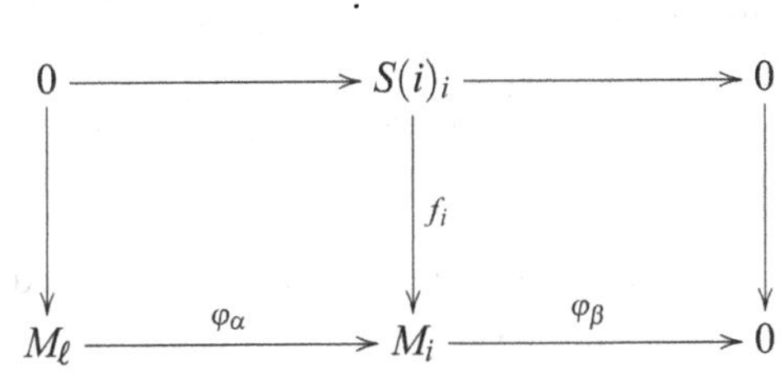

commutes, for all arrows $\ell \xrightarrow{\alpha} i$ and $i \xrightarrow{\beta} j$ in Q. Since f is injective this shows that $S(i)$ is a subrepresentation of M, and therefore, either $M \cong S(i)$ or M is not simple. □

Remark 2.10. Proposition 2.9 does not hold if the quiver has oriented cycles. For example, if Q is the quiver $\alpha \circlearrowleft 1$ then for each $\lambda \in k$, there is a simple representation $f_\lambda \circlearrowleft k$, where f_λ is given by multiplication by λ.

The vector space at vertex i of any representation can be described as a space of morphisms using the projective representation $P(i)$ as follows:

Theorem 2.11. *Let $M = (M_i, \psi_\alpha)$ be a representation of Q. Then, for any vertex i in Q, there is an isomorphism of vector spaces:*

$$Hom(P(i), M) \cong M_i.$$

Proof. Let $e_i = (i||i)$ be the constant path at i. Then $\{e_i\}$ is a basis of the vector space $P(i)_i$. Define a map

$$\begin{aligned} \phi : \mathrm{Hom}(P(i), M) &\longrightarrow M_i \\ f = (f_j)_{j \in Q_0} &\longmapsto f_i(e_i). \end{aligned}$$

If f is a morphism from $P(i)$ to M, then its component f_i is a linear map from $P(i)_i$ to M_i, which shows that the map ϕ is well defined, since $e_i \in P(i)_i$.

We will show that ϕ is an isomorphism of vector spaces. Let us use the notation $P(i) = (P(i)_j, \varphi_\alpha)$. First, we show that ϕ is linear. If $f, g \in \mathrm{Hom}(P(i), M)$ are two morphisms, then $\phi(f+g) = (f+g)_i(e_i) = f_i(e_i) + g_i(e_i) = \phi(f) + \phi(g)$, and if $\lambda \in k$ then $\phi(\lambda f) = (\lambda f)_i(e_i) = \lambda f_i(e_i) = \lambda\phi(f)$.

Next, we show that ϕ is injective. If $0 = \phi(f) = f_i(e_i)$, then the linear map f_i sends the basis $\{e_i\}$ to zero, and thus f_i is the zero map. We will now show that $f_j : P(i)_j \to M_j$ is the zero map, for any vertex j, and this will show that ϕ is injective. By definition of $P(i)$, the vector space $P(i)_j$ has a basis consisting of all paths from i to j. Let $c = (i|\alpha_1, \ldots, \alpha_t|j)$ be such a basis element, and consider the maps $\varphi_c = \varphi_{\alpha_t} \circ \cdots \circ \varphi_{\alpha_1}$ and $\varphi'_c = \varphi'_{\alpha_t} \circ \cdots \circ \varphi'_{\alpha_1}$ defined as the composition of the maps along the path c of the representation $P(i)$ and M, respectively. It follows from the definition of $P(i)$ that $\varphi_c(e_i) = c$. Since f is a morphism of representations, we have $f_j \varphi_c = \varphi'_c f_i$, and, since $f_i(e_i) = 0$, this implies that f_j maps c to zero. As c is an arbitrary basis element of $P(i)_j$, it follows that $f_j = 0$.

It remains to show that ϕ is surjective. Let $m_i \in M_i$. We want to construct a morphism $f : P(i) \to M$ such that $f_i(e_i) = m_i$. Let us start by fixing its component $f_i : P(i)_i \to M_i$ by requiring the condition we need, that is, $f_i(e_i) = m_i$. Since $\{e_i\}$ is a basis of $P(i)_i$, this condition defines the linear map f_i in a unique way. We can extend the map f_i to a morphism $f = (f_j)_{j \in Q_0}$ by *following the paths* in Q. More precisely, for any path c from i to a vertex j in Q, put $f_j(c) = \varphi'_c(m_i)$. This defines each map f_j on a basis of $P(i)_j$, and we extend this map linearly to the whole vector space $P(i)_j$. It follows from our construction that f is a morphism of representations, thus $f \in \mathrm{Hom}(P(i), M)$ and $\phi(f) = m_i$, so ϕ is surjective. □

Note that in the proof of Theorem 2.11, the particular structure of the representation $P(i)$ is the essential ingredient to show the injectivity and surjectivity of the map ϕ.

As an immediate consequence of Theorem 2.11, we can describe the morphisms between projective representations as follows:

Corollary 2.12. *Let i and j be vertices in Q.*

(1) The vector space $Hom(P(i), P(j))$ has a basis consisting of all paths from j to i in Q. In particular,

$$End\,(P(i)) = Hom(P(i), P(i)) \cong k.$$

(2) If $A = \oplus_{i \in Q_0} P(i)$, then the vector space $End\,(A) = Hom(A, A)$ has a basis consisting of all paths in Q.

Remark 2.13. In Chap. 4, we will see that the so-called path algebra of the quiver Q is isomorphic to End (A) as a vector space and as an algebra.

Proof. Theorem 2.11 implies that $\mathrm{Hom}(P(i), P(j))$ is isomorphic to $P(j)_i$, and this vector space has a basis consisting of all paths from j to i in Q. The fact that Q has no oriented cycles implies that End $(P(i))$ is of dimension one; hence End $(P(i)) \cong k$. This proves (1), and (2) is a direct consequence. □

Corollary 2.14. *The representation $P(j)$ is a simple representation if and only if $Hom(P(i), P(j)) = 0$ for all $i \neq j$.*

Proof. The representation $P(j)$ is simple if and only if j is a sink, which means that there are no paths from j to any other vertex i. The statement now follows from Corollary 2.12. □

2.2 Projective Resolutions and Radicals of Projectives

We have seen above that the projective representations can be used to describe the vector spaces of an arbitrary representation using the Hom functor. Now we will introduce another way of describing arbitrary representations by means of projective representations: the projective resolutions. As usual, there is a dual notion, the injective resolutions.

Definition 2.3. Let M be a representation of Q.

(1) A **projective resolution** of M is an exact sequence

$$\cdots \longrightarrow P_3 \longrightarrow P_2 \longrightarrow P_1 \longrightarrow P_0 \longrightarrow M \longrightarrow 0,$$

where each P_i is a projective representation.

(2) An **injective resolution** of M is an exact sequence

$$0 \longrightarrow M \longrightarrow I_0 \longrightarrow I_1 \longrightarrow I_2 \longrightarrow I_3 \longrightarrow \cdots,$$

where each I_i is an injective representation.

Theorem 2.15. *Let M be a representation of Q.*

(1) There exists a projective resolution of M of the form

$$0 \longrightarrow P_1 \longrightarrow P_0 \longrightarrow M \longrightarrow 0.$$

(2) There exists an injective resolution of M of the form

$$0 \longrightarrow M \longrightarrow I_0 \longrightarrow I_1 \longrightarrow 0.$$

Proof. We will show only (1), and to achieve this, we will construct the so-called standard projective resolution of M. Let $M = (M_i, \varphi_i)$, and denote by d_i the dimension of M_i. Define

$$P_1 = \bigoplus_{\alpha \in Q_1} d_{s(\alpha)} P(t(\alpha)) \qquad P_0 = \bigoplus_{i \in Q_0} d_i P(i),$$

where $d_i P(i)$ stands for the direct sum of d_i copies of $P(i)$.

Before defining the morphisms of the projective resolution, let us examine the representations P_0 and P_1. For every vector space M_i, we have $d_i = \dim M_i$ copies of $P(i)$ in P_0. The natural map g from P_0 to M will send the d_i copies of the constant path e_i in P_0 to a basis of M_i. Now in each copy of $P(i)$, the kernel of the map g contains a copy of $P(t(\alpha))$ for every α that starts at i. So, for every arrow α with $s(\alpha) = i$, we have $d_{s(\alpha)}$ copies of $P(t(\alpha))$ in the kernel of g, which justifies the definition of P_1.

To define the morphisms of the projective resolution, we introduce specific bases for each of the representations P_1, P_0 and M as follows: For each $i \in Q_0$, let $\{m_{i1}, \ldots, m_{id_i}\}$ be a basis for M_i, and thus

$$B'' = \{m_{ij} \mid i \in Q_0, j = 1, 2, \ldots, d_i\}$$

is a basis for M. Taking the standard bases for the projective representations, the set

$$B = \{c_{ij} \mid i \in Q_0, c_i \text{ a path with } s(c_i) = i, \text{ and } j = 1, \ldots, d_i\}$$

is a basis for P_0; and the set

$$B' = \{b_{\alpha j} \mid \alpha \in Q_1, b_\alpha \text{ a path with } s(b_\alpha) = t(\alpha), \text{ and } j = 1, \ldots, d_{s(\alpha)}\}$$

is a basis for P_1. Define a map g on the basis B by

$$g(c_{ij}) = \varphi_{c_i}(m_{ij}) \in M_{t(c_i)}$$

and extend g linearly to P_1. Define the map f on the basis B' by

$$f(b_{\alpha j}) = (\alpha b_\alpha)_j - b_\alpha^M,$$

where αb_α is the path from $s(\alpha)$ to $t(b_\alpha)$ given by the composition of α and b_α, and $b_\alpha^M = \sum_{\ell=1}^{d_{t(\alpha)}} \theta_\ell b_{\alpha\ell}$, where the θ_ℓ are the scalars that occur when writing $\varphi_\alpha(m_{s(\alpha)j})$ in the basis $\{m_{t(\alpha)\ell} \mid \ell = 1, \ldots, d_{t(\alpha)}\}$ of $M_{t(\alpha)}$; thus

$$\varphi_\alpha(m_{s(\alpha)j}) = \sum_{\ell=1}^{d_{t(\alpha)}} \theta_\ell m_{t(\alpha)\ell}. \tag{2.2}$$

We will now prove that the sequence

$$0 \longrightarrow \bigoplus_{\alpha \in Q_1} d_{s(\alpha)} P(t(\alpha)) \xrightarrow{f} \bigoplus_{i \in Q_0} d_i P(i) \xrightarrow{g} M \longrightarrow 0 \tag{2.3}$$

is exact.

g is surjective, because for any basis vector m_{ij} of M, we have $m_{ij} = g(e_{ij})$, where e_i is the constant path at vertex i.

ker $g \supset$ *im f*: It suffices to show that $g \circ f(b_{\alpha j}) = 0$ for any $b_{\alpha j}$ in the basis B'. We compute

$$\begin{aligned} g(f(b_{\alpha j})) &= g((\alpha b_\alpha)_j - b_\alpha^M) \\ &= \varphi_{\alpha b_\alpha}(m_{s(\alpha)j}) - \varphi_{b_\alpha}(\textstyle\sum_\ell \theta_\ell m_{t(\alpha)\ell}) \\ &= \varphi_{b_\alpha}\Big(\varphi_\alpha(m_{s(\alpha)j}) - \textstyle\sum_\ell \theta_\ell m_{t(\alpha)\ell}\Big) \\ &= \varphi_{b_\alpha}(0) \\ &= 0, \end{aligned}$$

where the next to last equation follows from (2.2).

ker $g \subset$ *im f*: First note that any $x \in \bigoplus_{i \in Q_0} d_i P(i)$ can be written as a linear combination of the basis B; thus

$$x = \sum_{c_{ij} \in B} \lambda_{c_{ij}} c_{ij} = x_0 + \sum_{c_{ij} \in B \setminus B_0} \lambda_{c_{ij}} c_{ij},$$

where B_0 is the subset of B consisting of constant paths (together with a choice of j),

$$B_0 = \{e_{ij} \mid i \in Q_0, j = 1, \ldots, d_i\},$$

and $x_0 = \sum_{e_{ij} \in B_0} \lambda_{e_{ij}} e_{ij}$. Any nonconstant path is the product of an arrow and another path; thus

$$x = x_0 + \sum_{c_{ij}: c_i = \alpha b_\alpha} \lambda_{c_{ij}} (\alpha b_\alpha)_j,$$

and using the definition of f, we get

$$x = x_0 + \sum_{c_{ij}:c_i=\alpha b_\alpha} \lambda_{c_{ij}} f(b_{\alpha j}) + \lambda_{c_{ij}} b_\alpha^M. \tag{2.4}$$

Let $x_1 = x_0 + \sum_{c_i=\alpha b_\alpha} \lambda_{c_{ij}} b_\alpha^M$. Note that $x - x_1 \in \operatorname{im} f$.

Define the degree of a linear combination of paths to be the length of the longest path that appears in it with nonzero coefficient. Note that $\deg x_1 < \deg x$ and $\deg x_0 = 0$.

Now let us suppose that $x \in \ker g$. We want to show that $x \in \operatorname{im} f$. Using (2.4) and the fact that $g \circ f = 0$, we get

$$0 = g(x) = g(x_1).$$

Summarizing, we have $x_1 \in \ker g$, $\deg x_1 < \deg x$ and $x - x_1 \in \operatorname{im} f$.

Now we repeat the argument with x_1 instead of x. We get $x_2 \in \ker g$, $\deg x_2 < \deg x_1 < \deg x$, and $x - x_2 \in \operatorname{im} f$. Continuing like this, we will eventually get $x_h \in \ker g$, $x - x_h \in \operatorname{im} f$, and $\deg x_h = 0$; thus x_h is a linear combination of constant paths, say $x_h = \sum_{i,j} \mu_{ij} e_{ij}$, for some $\mu_{ij} \in k$. By definition of g, we have

$$0 = g(x_h) = \sum_{i,j} \mu_{ij} m_{ij},$$

and since the m_{ij} form a basis of M, this implies that all μ_{ij} are zero. Hence $x_h = 0$ and thus $x \in \operatorname{im} f$.

f is injective. Suppose that

$$0 = f\left(\sum \lambda_{b_{\alpha h}} b_{\alpha h}\right) = \sum \lambda_{b_{\alpha h}} \left((\alpha b_\alpha)_h - b_\alpha^M\right).$$

Then

$$\sum \lambda_{b_{\alpha h}} (\alpha b_\alpha)_h = \sum \lambda_{b_{\alpha h}} b_\alpha^M = \sum \lambda_{b_{\alpha h}} \sum \theta_\ell b_{\alpha \ell}.$$

Let i_0 be a source in M, that is, i_0 is such that there is no arrow $j \to i_0$ with $d_j \neq 0$. Note that such an i_0 exists since Q has no oriented cycles. Then, since each of the paths b_α starts at the endpoint of the arrow α, none of the paths b_α can go through i_0, and this shows that $\lambda_{b_{\alpha h}}$ must be zero, for all arrows α with $s(\alpha) = i_0$. Now let i_1 be a source in $M \setminus i_0$, that is, i_1 is such that there is no arrow $j \to i_1$ with $j \neq i_0$ and $d_j \neq 0$. Since $\lambda_{b_{\alpha h}} = 0$ for all arrows α with $s(\alpha) = i_0$, there is no path $b_{\alpha h}$ with $\lambda_{b_{\alpha h}} \neq 0$ that goes through i_1. Continuing in this way, we show that every $\lambda_{b_{\alpha h}} = 0$, since Q has only finitely many arrows. Thus f is injective. This concludes the proof of the exactness of the standard resolution (2.3). □

Remark 2.16. There are other projective resolutions than the standard resolution.

Example 2.5. Let Q be the quiver $1 \longrightarrow 2 \longleftarrow 3$ and consider the representations $M = S(3) = 3$ and $M' = {}_{1}\begin{smallmatrix}2\\3\end{smallmatrix}$. Then we have the standard projective resolutions:

$$0 \longrightarrow 2 \longrightarrow \begin{smallmatrix}3\\2\end{smallmatrix} \longrightarrow 3 \longrightarrow 0$$

$$0 \longrightarrow 2 \oplus 2 \longrightarrow \begin{smallmatrix}1\\2\end{smallmatrix} \oplus \begin{smallmatrix}3\\2\end{smallmatrix} \oplus 2 \longrightarrow \begin{smallmatrix}1\,3\\2\end{smallmatrix} \longrightarrow 0.$$

The second resolution is not minimal in the sense that one can eliminate a direct summand $S(2) = 2$ in each of the projective modules and still have a projective resolution:

$$0 \longrightarrow 2 \longrightarrow \begin{smallmatrix}1\\2\end{smallmatrix} \oplus \begin{smallmatrix}3\\2\end{smallmatrix} \longrightarrow \begin{smallmatrix}1\,3\\2\end{smallmatrix} \longrightarrow 0.$$

Example 2.6. Let Q be the quiver

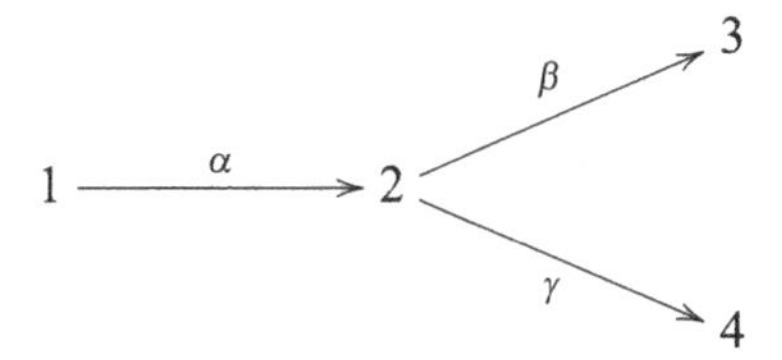

and consider the representation $M = \begin{smallmatrix}1\\2\,2\\3\,4\end{smallmatrix}$ given by

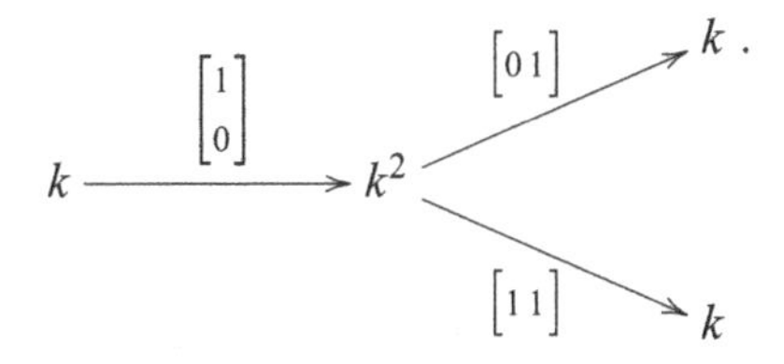

Then we have the standard projective resolution:

$$0 \to \begin{smallmatrix}2\\3\,4\end{smallmatrix} \oplus (3 \oplus 3) \oplus (4 \oplus 4) \to \begin{smallmatrix}1\\2\\3\,4\end{smallmatrix} \oplus \left(\begin{smallmatrix}2\\3\,4\end{smallmatrix} \oplus \begin{smallmatrix}2\\3\,4\end{smallmatrix}\right) \oplus 3 \oplus 4 \to \begin{smallmatrix}1\\2\,2\\3\,4\end{smallmatrix} \to 0.$$

Again, this resolution is not minimal in the sense that one can eliminate three direct summands in each of the projective modules and still have a projective resolution:

$$0 \longrightarrow 3 \oplus 4 \longrightarrow \begin{smallmatrix} 1 \\ 2 \\ 3\,4 \end{smallmatrix} \oplus \begin{smallmatrix} 2 \\ 3\,4 \end{smallmatrix} \longrightarrow \begin{smallmatrix} 1 \\ 2\,2 \\ 3\;\,4 \end{smallmatrix} \longrightarrow 0.$$

To make this notion of minimality more precise, we need the definitions of projective covers and injective envelopes.

Definition 2.4. Let $M \in \operatorname{rep} Q$. A **projective cover** of M is a projective representation P together with a surjective morphism $g \colon P \to M$ with the property that, whenever $g' \colon P' \to M$ is a surjective morphism with P' projective, then there exists a *surjective* morphism $h : P' \twoheadrightarrow P$ such that the diagram

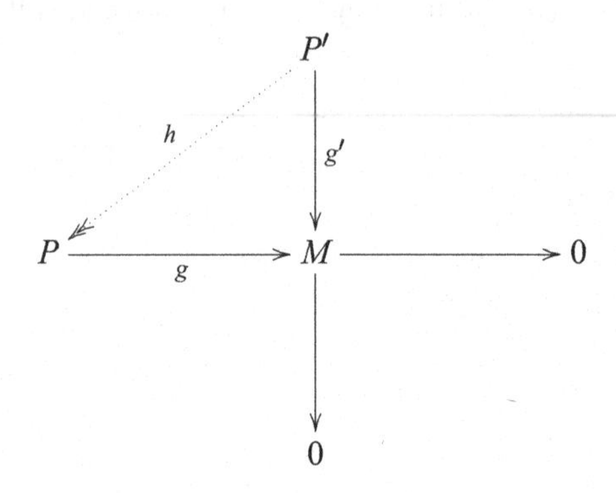

commutes, that is, $gh = g'$.

An **injective envelope** of M is an injective representation I together with an injective morphism $f \colon M \to I$ with the property that, whenever $f' \colon M \to I'$ is an injective morphism into an injective representation I', then there exists an *injective* morphism $h : I \hookrightarrow I'$ such that the diagram

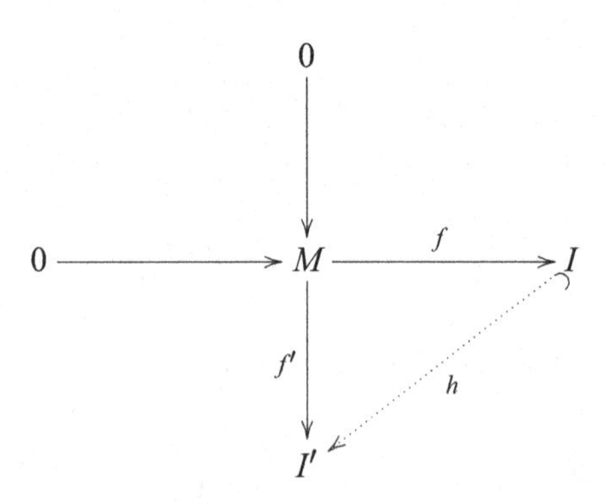

commutes, that is, $hf = f'$.

Definition 2.5. A projective resolution

$$\cdots \longrightarrow P_3 \xrightarrow{f_3} P_2 \xrightarrow{f_2} P_1 \xrightarrow{f_1} P_0 \xrightarrow{f_0} M \longrightarrow 0$$

is called **minimal** if $f_0\colon P_0 \to M$ is a projective cover and $f_i\colon P_i \to \ker f_{i-1}$ is a projective cover, for every $i > 0$.

An injective resolution

$$0 \longrightarrow M \xrightarrow{f_0} I_0 \xrightarrow{f_1} I_1 \xrightarrow{f_2} I_2 \xrightarrow{f_3} I_3 \longrightarrow \cdots$$

is called **minimal** if $f_0\colon M \to I_0$ is an injective envelope and, for every $i > 0$, $f_i\colon \operatorname{coker} f_{i-1} \to I_i$ is an injective envelope.

The next two propositions show that projective covers are unique up to isomorphism.

Proposition 2.17. *Let $g\colon P \to M$ be a projective cover of M and let $g'\colon P' \to M$ be a surjective morphism with P' projective. Then P is isomorphic to a direct summand of P'.*

Proof. From the definition of projective covers, we see that there exists a surjective morphism $h\colon P' \to P$. This morphism gives rise to an exact sequence:

$$0 \longrightarrow \ker h \longrightarrow P' \xrightarrow{h} P \longrightarrow 0\,.$$

Since P is projective, Corollary 2.4 implies that this sequence splits, and then the result follows from Proposition 1.8. □

Proposition 2.18. *Let $g\colon P \to M$ and $g'\colon P' \to M$ be projective covers of M. Then P is isomorphic to P'.*

Proof. From Proposition 2.17, we conclude that P is isomorphic to a direct summand of P', and P' is isomorphic to a direct summand of P. Thus $P \cong P'$. □

Remark 2.19. The dual statements to the propositions 2.17 and 2.18 about injective envelopes hold too. We leave the statements and their proofs as an exercise.

We introduce now the concept of a free representation. The prototype of a free representation is the direct sum of the indecomposable projective representations, and in general, free representations are direct sums of this prototype.

Definition 2.6. Let $A = \oplus_{i \in Q_0} P(i)$. A representation $F \in \operatorname{rep} Q$ is called **free** if $F \cong A \oplus \cdots \oplus A$.

Proposition 2.20. *A representation $M \in \operatorname{rep} Q$ is projective if and only if there exists a free representation $F \in \operatorname{rep} Q$ such that M is isomorphic to a direct summand of F.*

Proof.

($\Leftarrow$) By the Krull–Schmidt Theorem 1.2 and Proposition 2.8, every direct summand of F is a direct sum of $P(i)$'s, hence projective, by Proposition 2.7.

($\Rightarrow$) Suppose that M is projective and $\underline{\dim}\, M = (d_i)_{i \in Q_0}$. The standard projective resolution of M gives a surjective morphism $g: \oplus\, d_i\, P(i) \to M$. Thus there is a short exact sequence:

$$0 \longrightarrow \ker g \longrightarrow \oplus d_i P(i) \xrightarrow{g} M \longrightarrow 0$$

Since M is projective, this sequence splits, and therefore M is isomorphic to a direct summand $\oplus\, d_i\, P(i)$. □

Corollary 2.21. *Any projective representation $P \in$ rep Q is a direct sum of $P(i)$'s, that is,*

$$P \cong P(i_1) \oplus \cdots \oplus P(i_t),$$

with $i_1, \ldots, i_t$ not necessarily distinct.

Proof. This follows directly from Proposition 2.20 □

Our next goal is to show that, in rep Q, subrepresentations of projective representations are projective. We start by introducing a particular subrepresentation of $P(i)$.

Definition 2.7. Let $P(i) = (P(i)_j, \varphi_\alpha)$ be the projective representation at vertex i. The **radical** of $P(i)$ is the representation rad $P(i) = (R_j, \varphi'_\alpha)$ defined by

$$R_i = 0, \quad R_j = P(i)_j \text{ if } i \neq j, \quad \text{and} \quad \varphi'_\alpha = \begin{cases} 0 & \text{if } s(\alpha) = i \\ \varphi_\alpha & \text{otherwise.} \end{cases}$$

The next lemma shows that the radical of $P(i)$ is the maximal proper subrepresentation of $P(i)$.

Lemma 2.22. *Any proper subrepresentation of $P(i)$ is contained in rad $P(i)$.*

Proof. Suppose $f: M \hookrightarrow P(i)$ is an injective morphism of representations. Let $M = (M_i, \psi_\alpha)$ and $P(i) = (P(i)_j, \varphi_\alpha)$. It is clear that if $M_i = 0$, we have that $f(M_i) \subset \text{rad}\, P(i)$, so let us suppose that $M_i \neq 0$. We will show that this implies that the morphism f is an isomorphism. Since $P(i)_i \cong k$, it follows that $M_i \cong k$, and there is an element $m_i \in M_i$ such that $f_i(m_i) = e_i$. Now let j be any vertex, and let c be a path from i to j. Then

$$c = \varphi_c(e_i) = \varphi_c(f_i(m_i)) = f_j(\psi_c(m_i)) \in \text{im}\, f_j,$$

where the first identity is shown in Remark 2.1, and the third identity holds because f is a morphism of representations. Thus we see that the arbitrary element c of the basis of $P(i)_j$ lies in the image of f_j, which implies that f is surjective, hence an isomorphism, and so M is not a proper subrepresentation of $P(i)$. □

Lemma 2.23. *If $P(i)$ is simple, then rad $P(i) = 0$. If $P(i)$ is not simple, then the radical of $P(i)$ is projective.*

Proof. We will show that rad $P(i)$ is isomorphic to the projective representation $P = \oplus_{\alpha:s(\alpha)=i} P(t(\alpha))$. If $i \neq j$, then $(\text{rad}\, P(i))_j = P(i)_j$ has as a basis the set of paths from i to j. Define a morphism $f = (f_j)_{j\in Q_0} : \text{rad}\, P(i) \to P$ on this basis by

$$f_j(i|\alpha, \beta_1, \dots \beta_s|j) = (t(\alpha)|\beta_1, \dots, \beta_s|j).$$

Then f_j sends the basis of $(\text{rad}\, P(i))_j$ to a basis of P_j, for each $j \in Q_0$, and thus f is an isomorphism. □

Theorem 2.24. *Subrepresentations of projective representations in rep Q are projective.*

Remark 2.25. The subrepresentation inherits the projectivity. Categories with this property are called **hereditary**.

Proof. Suppose that P is a projective representation with dimension vector $(d_i)_{i\in Q_0}$. We will prove the theorem by induction on $d = \sum_{i\in Q_0} d_i$, the dimension of P.

If $d = 1$, then P is simple and there is nothing to prove. So suppose that $d > 1$. Let M be a subrepresentation of P and let $u : M \to P$ be the inclusion morphism. By Corollary 2.21, we have $P \cong P(i_1) \oplus \cdots \oplus P(i_t)$ for some vertices $i_1, \dots, i_t$, and thus the inclusion u is of the form

$$u = \begin{bmatrix} u_1 \\ \vdots \\ u_t \end{bmatrix}$$

with $\text{im}\, u_j \subset P(i_j)$. It follows that $M \cong \text{im}\, u_1 \oplus \cdots \oplus \text{im}\, u_t$, and, by Proposition 2.7, it suffices to show that $\text{im}\, u_j$ is projective for each j. This is obvious in the case where $\text{im}\, u_j = P(i_j)$, so let us suppose that $\text{im}\, u_j$ is a proper subrepresentation of $P(i_j)$. Then $\text{im}\, u_j$ is a subrepresentation of rad $P(i_j)$, and rad $P(i_j)$ is projective, by Lemma 2.23. Moreover, the dimension of rad $P(i_j)$ is strictly smaller than d, and, by induction, we conclude that $\text{im}\, u_j$ is projective, which completes the proof. □

As a consequence of Theorem 2.24, we obtain the following result on morphisms into projective modules:

Corollary 2.26. *Let $f: M \to P$ be a nonzero morphism from an indecomposable representation M to a projective representation P. Then M is projective, and f is injective.*

Proof. Since the image of f is a subrepresentation of P, it is projective, by Theorem 2.24. Therefore, the short exact sequence

$$0 \longrightarrow \ker f \longrightarrow M \longrightarrow \operatorname{im} f \longrightarrow 0$$

splits, and then Proposition 1.8 implies that $\operatorname{im} f$ is isomorphic to a direct summand of M. But M is indecomposable, so $M \cong \operatorname{im} f$ is projective and $\ker f = 0$. □

Corollary 2.26 shows that when we construct the Auslander–Reiten quiver of Q, we must start with the projective representations and that the projective representations are partially ordered by inclusion.

2.3 Auslander–Reiten Translation

In this section, we will define the Auslander–Reiten translation, which is fundamental for the Auslander–Reiten theory and Auslander–Reiten quivers. We consider at this point only the Auslander–Reiten translation in the category of quiver representations. Later, we will also consider the more general situation of bound quiver representations and modules.

We start with another notion from category theory.

Categories 5 *If $\mathscr{C}, \mathscr{D}$ are two categories. We say that two functors $F_1, F_2 : \mathscr{C} \to \mathscr{D}$ are* functorially isomorphic, *and we write $F_1 \cong F_2$, if for every object $M \in \mathscr{C}$, there exists an isomorphism $\eta_M : F_1(M) \to F_2(M) \in \mathscr{D}$ such that, for every morphism $f : M \to N$ in $\mathscr{C}$, the following diagram commutes:*

$$\begin{array}{ccc} F_1(M) & \xrightarrow{F_1(f)} & F_1(N) \\ \downarrow{\scriptstyle \eta_M} & & \downarrow{\scriptstyle \eta_N} \\ F_2(M) & \xrightarrow{F_2(f)} & F_2(N) \end{array}$$

A covariant functor $F : \mathscr{C} \to \mathscr{D}$ is called an **equivalence of categories** *if there exists a functor $G : \mathscr{D} \to \mathscr{C}$ such that $G \circ F \cong 1_{\mathscr{C}}$ and $F \circ G \cong 1_{\mathscr{D}}$. The functor G is called a* quasi-inverse *functor for F.*

A contravariant functor F that has a (contravariant) quasi-inverse is called a **duality**

2.3.1 Duality

Let Q be a quiver without oriented cycles, and let $Q^{\rm op}$ be the quiver obtained from Q by reversing each arrow. Thus $Q_0^{\rm op} = Q_0$ and $Q_1^{\rm op} = \{\alpha^{\rm op} \mid \alpha \in Q_1\}$ with $s(\alpha^{\rm op}) = t(\alpha)$ and $t(\alpha^{\rm op}) = s(\alpha)$.

In this section, we need to work with projective and injective representations of both Q and $Q^{\rm op}$. To distinguish between these, we will often use the notation $P_Q(i), I_Q(i)$ for the representations of Q and $P_{Q^{\rm op}}(i), I_{Q^{\rm op}}(i)$ for the representations of $Q^{\rm op}$.

The **duality**

$$D = \mathrm{Hom}_k(-, k) : \mathrm{rep}\, Q \longrightarrow \mathrm{rep}\, Q^{\rm op}$$

is the contravariant functor defined as follows:

- On objects $M = (M_i, \varphi_\alpha)$, we have

  $$DM = (DM_i, D\varphi_{\alpha^{\rm op}})_{i \in Q_0, \alpha \in Q_1},$$

  where DM_i is the dual vector space of the vector space M_i, and thus $DM_i = \mathrm{Hom}_k(M_i, k)$ is the space of linear maps $M_i \to k$; and if α is an arrow in Q then $D\varphi_{\alpha^{\rm op}}$ is the pullback of φ_α, and thus

  $$\begin{aligned} D\varphi_{\alpha^{\rm op}} : DM_{t(\alpha)} &\longrightarrow DM_{s(\alpha)} \\ u &\longmapsto u \circ \varphi_\alpha . \end{aligned}$$

- On morphisms $f : M \to N$ in $\mathrm{rep}\, Q$, we have $Df : DN \to DM$ in $\mathrm{rep}\, Q^{\rm op}$ defined by $Df(u) = u \circ f$:

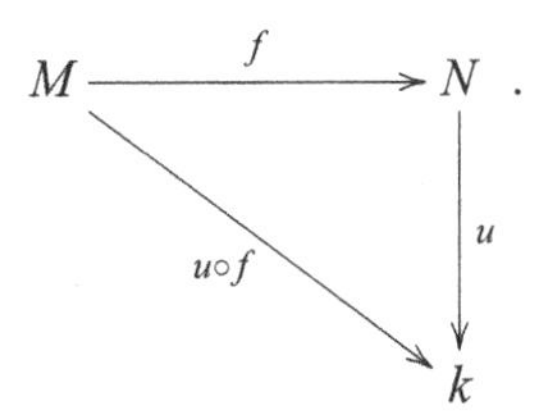

If we compose the duality of Q with the duality of $Q^{\rm op}$, we get the identity functor $1_{\mathrm{rep}\, Q}$; thus the quasi-inverse of D_Q is $D_{Q^{\rm op}}$.

Let $\mathrm{proj}\, Q$ be the category of projective representations of Q and let $\mathrm{inj}\, Q$ be category of injective representations of Q. Thus the objects in $\mathrm{proj}\, Q$ are the projective representations of Q, and the morphisms are the morphisms between projective representations.

Proposition 2.27. *We have $D(P_Q(i)) = I_{Q^{\rm op}}(i)$, for all vertices $i \in Q_0$, in particular, the duality restricts to a duality $\mathrm{proj}\, Q \to \mathrm{inj}\, Q^{\rm op}$.*

Proof. Exercise 2.10. □

Example 2.7. Let Q be the quiver $1 \longrightarrow 2 \longleftarrow 3$ as in Example 1.14. The indecomposable representations of the subcategory proj Q are the three projective representations:

$$\begin{smallmatrix}1\\2\end{smallmatrix} \quad 2 \quad \begin{smallmatrix}3\\2\end{smallmatrix}.$$

The quiver Q^{op} is $1 \longleftarrow 2 \longrightarrow 3$, and the indecomposable representations of the subcategory inj Q^{op} are

$$\begin{smallmatrix}2\\1\end{smallmatrix} \quad 2 \quad \begin{smallmatrix}2\\3\end{smallmatrix}.$$

2.3.2 Nakayama Functor

Let A be the free representation given as the direct sum of the indecomposable projective representations of Q, that is, $A = \oplus_{j \in Q_0} P(j)$.

Consider the contravariant functor $\mathrm{Hom}(-, A)$. We know already from Sect. 1.4 that the Hom functors map representations of Q to vector spaces, and thus $\mathrm{Hom}(X, Y)$ is a vector space for every pair of representations X, Y. But now, instead of the arbitrary representation Y, we use the special free representation A, and, in this case, we can give $\mathrm{Hom}(X, A)$ the structure of a representation $(M_i, \varphi_{\alpha^{\mathrm{op}}})$ of the opposite quiver Q^{op} as follows: Define the vector space at vertex i as $M_i = \mathrm{Hom}(X, P(i))$ for every $i \in Q_0$, and for an arrow α from i to j in Q, define a linear map $\varphi_{\alpha^{\mathrm{op}}} : \mathrm{Hom}(X, P(j)) \to \mathrm{Hom}(X, P(i))$ as $\varphi_{\alpha^{\mathrm{op}}}(f) = \alpha \circ f$; thus we have the diagram

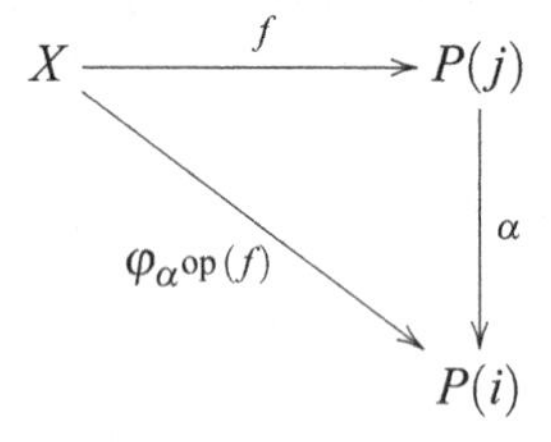

where we use the fact that α, being a path from i to j, gives a morphism from $P(j)$ to $P(i)$, by Corollary 2.12. Thus $\mathrm{Hom}(X, A)$ is a representation of Q^{op}.

To show that $\mathrm{Hom}(-,A)$ is a functor from rep Q to rep Q^{op}, we must check that the image under $\mathrm{Hom}(-,A)$ of any morphism $g: X \to X'$ of representations of Q is a morphism of representations of Q^{op}. This means we must check for every arrow $i \xrightarrow{\alpha} j$ in Q that the following diagram commutes:

$$\begin{array}{ccc} \mathrm{Hom}(X',P(j)) & \xrightarrow{\varphi'_{\alpha^{\mathrm{op}}}} & \mathrm{Hom}(X',P(i)) \\ \Big\downarrow {\scriptstyle g^*=\mathrm{Hom}(g,P(j))} & & \Big\downarrow {\scriptstyle g^*=\mathrm{Hom}(g,P(i))} \\ \mathrm{Hom}(X,P(j)) & \xrightarrow[\varphi_{\alpha^{\mathrm{op}}}]{} & \mathrm{Hom}(X,P(i)). \end{array}$$

To do so, let $f \in \mathrm{Hom}(X', P(j))$, then $g^*\varphi'_{\alpha^{\mathrm{op}}}(f) = g^*(\alpha \circ f) = (\alpha \circ f) \circ g$, whereas $\varphi_{\alpha^{\mathrm{op}}} \circ g^*(f) = \varphi_{\alpha^{\mathrm{op}}}(f \circ g) = \alpha \circ (f \circ g) = (\alpha \circ f) \circ g$, so the diagram commutes. We have shown the following:

Proposition 2.28. *$\mathrm{Hom}(-,A)$ is a functor from rep Q to rep Q^{op}.*

Composing the two contravariant functors D and $\mathrm{Hom}(-,A)$, we get the following important covariant functor:

Definition 2.8. The functor $\nu = D\mathrm{Hom}(-,A) : \mathrm{rep}\, Q \to \mathrm{rep}\, Q$ is called the **Nakayama functor**:

$$\mathrm{rep}\, Q \xrightarrow{\mathrm{Hom}(-,A)} \mathrm{rep}\, Q^{\mathrm{op}} \xrightarrow{D} \mathrm{rep}\, Q, \qquad \nu : \mathrm{rep}\, Q \to \mathrm{rep}\, Q$$

Corollary 2.26 implies that the functor $\mathrm{Hom}(-,A)$ is zero on all representations which have no projective direct summands. Therefore we must study the behavior of the Nakayama functor when applied to an indecomposable projective representation, say, $P_Q(i)$. Let us first consider the functor $\mathrm{Hom}(-,A)$ only. Let $M = \mathrm{Hom}(P_Q(i), A)$, and use the habitual notation $M = (M_j, \varphi_{\alpha^{\mathrm{op}}})_{j \in Q_0, \alpha \in Q_1}$ to denote the representation. Then the vector space M_j is $\mathrm{Hom}(P_Q(i), P_Q(j))$; thus, according to Corollary 2.14, the space M_j has a basis consisting of all paths from j to i in Q. In terms of Q^{op} this can be rephrased as M_j has a basis consisting of all paths from i to j in Q^{op}.

Moreover, for any arrow $h \xrightarrow{\alpha} j$ in Q, the linear map $\varphi_{\alpha^{\mathrm{op}}} : M_j \to M_h$ maps the basis element c, which is a path from i to j in Q^{op}, to the basis element $c\alpha^{\mathrm{op}}$, which is a path from i to h in Q^{op}. This shows that $M = \mathrm{Hom}(P_Q(i), A)$ is the indecomposable projective Q^{op}-representation $P_{Q^{\mathrm{op}}}(i)$ at vertex i.

Thus the restriction of $\mathrm{Hom}(-, A)$ to the subcategory proj Q gives a duality of categories proj $Q \to$ proj Q^{op}, whose quasi-inverse is given by $\mathrm{Hom}_{Q^{\mathrm{op}}}(-, A^{\mathrm{op}})$, where A^{op} is the sum of all indecomposable projective Q^{op}-representations.

To obtain the Nakayama functor ν, we must now form the composition with the duality D. Note that $DA^{\mathrm{op}} = \oplus_{i \in Q_0} I_Q(i)$.

Proposition 2.29. *The restriction of ν to proj Q is an equivalence of categories proj $Q \to$ inj Q whose quasi-inverse is given by*

$$\nu^{-1} = Hom(DA^{op}, -)\colon inj\ Q \longrightarrow proj\ Q.$$

Moreover, for any vertex i,

$$\nu P(i) = I(i),$$

and if c is a path from i to j, and $f_c \in Hom(P(j), P(i))$ is the corresponding morphism, then

$$\nu f_c : I(j) \to I(i)$$

is the morphism given by the cancellation of the path c.

Proof. The functor ν is an equivalence because it is the composition of the two dualities D and $\mathrm{Hom}(-, A)$. Its quasi-inverse ν^{-1} is the composition of the quasi-inverses of D and $\mathrm{Hom}(-, A)$, thus $\nu^{-1} = \mathrm{Hom}_{Q^{\mathrm{op}}}(-, A^{\mathrm{op}}) \circ D$. Note that since $\mathrm{Hom}_{Q^{\mathrm{op}}}(DX, DY) \cong \mathrm{Hom}_Q(Y, X)$ for all $X, Y \in \mathrm{rep}\, Q$, we have in particular that $\mathrm{Hom}_{Q^{\mathrm{op}}}(DX, A^{\mathrm{op}}) \cong \mathrm{Hom}_Q(DA^{\mathrm{op}}, X)$, whence $\nu^{-1} = \mathrm{Hom}(DA, -)$. Finally,

$$\nu P_Q(i) = D\mathrm{Hom}(P_Q(i), A) = D(P_{Q^{\mathrm{op}}}(i)) = I_Q(i).$$

To show the last statement, let c be a path from i to j, and let $f_c\colon P_Q(j) \to P_Q(i)$ be defined by $f(x) = cx$ as in the proposition. Let f_c^* be the image of f_c under the functor $\mathrm{Hom}(-, A)$. Thus $f_c^*\colon \mathrm{Hom}(P_Q(i), A) \to \mathrm{Hom}(P_Q(j), A)$ maps a morphism g to the pullback $g \circ f_c^*$. Now using $\mathrm{Hom}(P_Q(x), A) \cong P_{Q^{\mathrm{op}}}(x)$, we see that $f_c^*\colon P_{Q^{\mathrm{op}}}(i) \to P_{Q^{\mathrm{op}}}(j)$ is given by $f(y) = c^{\mathrm{op}}y$, where c^{op} denotes the opposite of the path c in Q^{op}. Finally, $\nu f_c = Df_c^*$ is the map sending $D(c^{\mathrm{op}}y)$ to $D(y)$; thus the result follows from $D(c^{\mathrm{op}}y) = D(y)c$. □

Example 2.8. Let Q be the quiver $1 \longrightarrow 2 \longrightarrow 3$. Its Auslander–Reiten quiver is

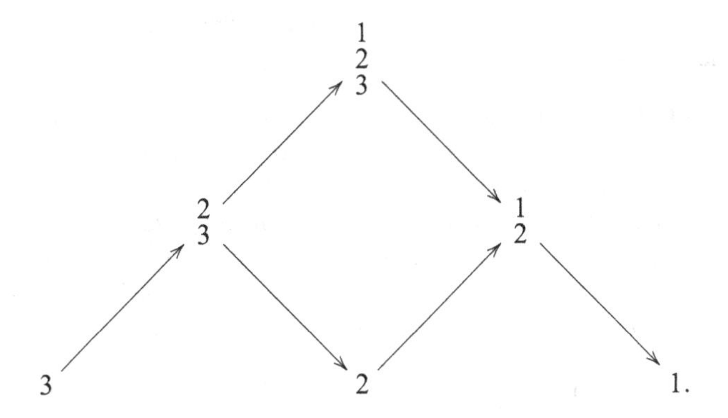

Note that the Nakayama functor ν sends proj Q to inj Q as follows:

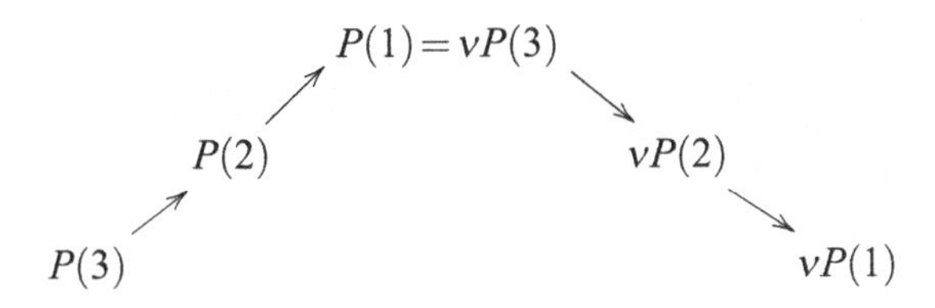

Example 2.9. Let Q be the quiver

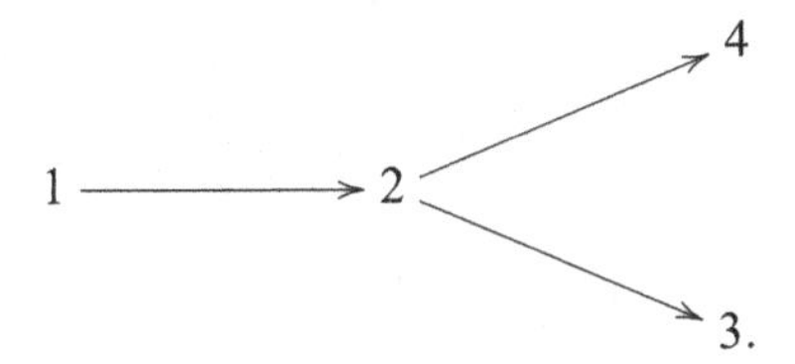

Then proj Q and inj Q are as follows:

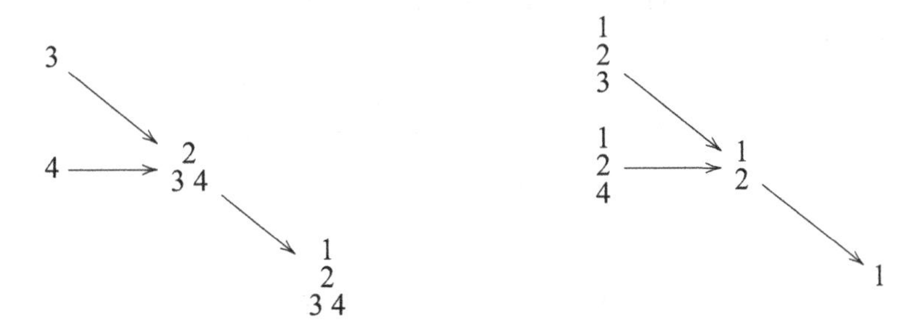

Before we can state the next property of the Nakayama functor, we need to introduce the notion of exactness for functors.

Categories 6 *Let $\mathscr{C}$ and $\mathscr{D}$ be abelian categories. A (covariant or contravariant) functor $F:\mathscr{C} \to \mathscr{D}$ is called* exact *if it maps exact sequences in $\mathscr{C}$ to exact sequences in $\mathscr{D}$. For example, every equivalence or duality of abelian categories is exact. Many nice functors, the Hom-functors for example, are not exact but have the weaker property of being left exact or right exact, which we define below. The definition of these notions is different for covariant and contravariant functors.*

Let $F:\mathscr{C} \to \mathscr{D}$ be a covariant functor. F is called *left exact* if for any exact sequence

$$0 \longrightarrow L \xrightarrow{f} M \xrightarrow{g} N$$

the sequence

$$0 \longrightarrow F(L) \xrightarrow{F(f)} F(M) \xrightarrow{F(g)} F(N)$$

is exact.
F is called *right exact* if for any exact sequence

$$L \xrightarrow{f} M \xrightarrow{g} N \longrightarrow 0$$

the sequence

$$F(L) \xrightarrow{F(f)} F(M) \xrightarrow{F(g)} F(N) \longrightarrow 0$$

is exact.

Let $G:\mathscr{C} \to \mathscr{D}$ be a contravariant functor. G is called *left exact* if for any exact sequence

$$L \xrightarrow{f} M \xrightarrow{g} N \longrightarrow 0$$

the sequence

$$0 \longrightarrow G(N) \xrightarrow{G(g)} G(M) \xrightarrow{G(f)} G(L)$$

is exact.

G is called *right exact* if for any exact sequence

$$0 \longrightarrow L \xrightarrow{f} M \xrightarrow{g} N$$

the sequence

$$G(N) \xrightarrow{G(g)} G(M) \xrightarrow{G(f)} G(L) \longrightarrow 0$$

is exact.

We have shown in Sect. 1.4 that the Hom functors $\text{Hom}(X, -)$ and $\text{Hom}(-, X)$ are left exact. Since the Nakayama functor is the composition of the left exact functor $\text{Hom}(-, A)$ and the exact contravariant functor D, we get the following proposition:

Proposition 2.30. *The Nakayama functor ν is right exact.*

Example 2.10. In the setting of Example 2.8 there is a short exact sequence

$$0 \longrightarrow 3 \xrightarrow{f} \begin{smallmatrix}1\\2\\3\end{smallmatrix} \xrightarrow{g} \begin{smallmatrix}1\\2\end{smallmatrix} \longrightarrow 0,$$

and applying ν yields the exact sequence

$$\begin{smallmatrix}1\\2\\3\end{smallmatrix} \xrightarrow{\nu f} 1 \xrightarrow{\nu g} 0 \longrightarrow 0.$$

This confirms that ν is right exact. Since the morphism νf is clearly not injective, this also shows that ν is not exact.

2.3.3 The Auslander–Reiten Translations τ, τ^{-1}

Let Q be a quiver without oriented cycles, and let M be an indecomposable representation of Q.

Definition 2.9. Let

$$0 \longrightarrow P_1 \xrightarrow{p_1} P_0 \xrightarrow{p_0} M \longrightarrow 0$$

be a minimal projective resolution. Applying the Nakayama functor, we get an exact sequence

$$0 \longrightarrow \tau M \longrightarrow \nu P_1 \xrightarrow{\nu p_1} \nu P_0 \xrightarrow{\nu p_0} \nu M \longrightarrow 0,$$

where $\tau M = \ker \nu p_1$ is called the Auslander–Reiten translate of M and τ the **Auslander–Reiten translation.**

Let

$$0 \longrightarrow M \xrightarrow{i_0} I_0 \xrightarrow{i_1} I_i \longrightarrow 0$$

be a minimal injective resolution. Applying the inverse Nakayama functor, we get an exact sequence

$$0 \longrightarrow \nu^{-1}M \xrightarrow{\nu^{-1}i_0} \nu^{-1}I_0 \xrightarrow{\nu^{-1}i_1} \nu^{-1}I_1 \longrightarrow \tau^{-1}M \longrightarrow 0,$$

where $\tau^{-1}M = \operatorname{coker} \nu^{-1}i_1$ is called the inverse Auslander–Reiten translate of M and τ^{-1} the inverse Auslander–Reiten translation.

Example 2.11. Continuing Example 2.10, we compute $\tau\,{}^{1}_{2}$. We have already constructed the minimal projective resolution and applied the Nakayama functor to it. It only remains to compute the kernel of νf; thus

$$\tau\,{}^{1}_{2} = \ker \nu f = {}^{2}_{3}.$$

Remark 2.31. The Auslander–Reiten translation has been introduced by Auslander and Reiten in [10].

2.4 Extensions and Ext

In this section, Q always denotes a quiver without oriented cycles. We give here a short account on the Ext^1-groups; for further information we refer to [53, Sect. 7.2].

Let $M \in \operatorname{rep} Q$ and take a projective resolution

$$0 \longrightarrow P_1 \xrightarrow{f} P_0 \xrightarrow{g} M \longrightarrow 0$$

of M in $\operatorname{rep} Q$. Thus P_0 and P_1 are projective representations and the above sequence is exact. Let N be any representation in $\operatorname{rep} Q$. Then we can apply the functor $\operatorname{Hom}(-, N)$ to this projective resolution, and as a result we get the exact sequence

$$0 \to \operatorname{Hom}(M,N) \xrightarrow{g^*} \operatorname{Hom}(P_0,N) \xrightarrow{f^*} \operatorname{Hom}(P_1,N) \to \operatorname{Ext}^1(M,N) \to 0,$$

where $\operatorname{Ext}^1(M,N) = \operatorname{coker} f^*$ is called the **first group of extensions** of M and N.

Remark 2.32. In arbitrary categories, projective resolutions do not necessarily stop after two steps; in fact, they might not even stop at all. Thus a projective resolution in a general category is of the form

$$\cdots \longrightarrow P_n \xrightarrow{f_n} P_{n-1} \longrightarrow \cdots \longrightarrow P_1 \xrightarrow{f_1} P_0 \xrightarrow{f_0} M \longrightarrow 0$$

and applying $\mathrm{Hom}(-,N)$ yields a so-called *cochain complex*

$$0 \longrightarrow \mathrm{Hom}(M,N) \xrightarrow{f_0^*} \mathrm{Hom}(P_0,N) \xrightarrow{f_1^*} \cdots \xrightarrow{f_n^*} \mathrm{Hom}(P_n,N) \longrightarrow \cdots,$$

which means that $f_i^* f_{i-1}^* = 0$, for all i. One then defines the ith extension group $\mathrm{Ext}^i(M,N)$ for $i \geq 1$ to be the ith cohomology group of this complex, that is,

$$\mathrm{Ext}^i(M,N) = \ker f_{i+1}^* / \mathrm{im}\, f_i^*.$$

One can show that this definition does not depend on the choice of the projective resolution; see, for example, [53, Proposition 6.4].

In the category rep Q, all the Ext^i-groups, with $i \geq 2$, vanish, because the minimal projective resolutions are of the form

$$0 \longrightarrow P_1 \xrightarrow{f} P_0 \xrightarrow{g} M \longrightarrow 0\,.$$

On the other hand, the Ext^1-groups provide very interesting information.

Our next goal is to show that the vector space $\mathrm{Ext}^1(M,N)$ is isomorphic to the vector space of extensions of M by N.

Definition 2.10. An **extension** ζ of M by N is a short exact sequence $0 \to N \xrightarrow{f} E \xrightarrow{g} M \to 0$. Two extensions ζ and ζ' are called *equivalent* if there is a commutative diagram:

$$\begin{array}{ccccccccc}
\zeta: 0 & \longrightarrow & N & \xrightarrow{f} & E & \xrightarrow{g} & M & \longrightarrow & 0 \\
 & & \downarrow = & & \downarrow \cong & & \downarrow = & & \\
\zeta': 0 & \longrightarrow & N & \xrightarrow{f'} & E' & \xrightarrow{g'} & M & \longrightarrow & 0
\end{array}$$

Example 2.12. Let Q be the quiver

$$1 \underset{\beta}{\overset{\alpha}{\rightrightarrows}} 2\,,$$

let $N = S(2)$, $M = S(1)$ be the two simple modules and let

$$E = k \overset{1}{\underset{0}{\rightrightarrows}} k \quad \text{and} \quad E' = k \overset{0}{\underset{1}{\rightrightarrows}} k\,.$$

Then the short exact sequences

$$\begin{array}{l} \zeta : 0 \longrightarrow S(2) \xrightarrow{f} E \xrightarrow{g} S(1) \longrightarrow 0 \\ \zeta' : 0 \longrightarrow S(2) \xrightarrow{f'} E' \xrightarrow{g'} S(1) \longrightarrow 0 \end{array}$$

are not equivalent, because E and E' are not isomorphic.

An extension is *split* if the short exact sequence is split, that is, if the extension is equivalent to the short exact sequence:

$$0 \longrightarrow N \longrightarrow N \oplus M \longrightarrow M \longrightarrow 0.$$

Given two extensions ζ and ζ' of M by N, we define their sum $\zeta + \zeta'$ as follows: Let $E'' = \{(x, x') \in E \times E' \mid g(x) = g'(x')\}$ be the so-called *pull back* of g and g', and define F to be the quotient of E'' by the subspace $\{(f(n), -f'(n)) \in E \oplus E' \mid n \in N\}$, compare with Exercises 1.8 and 1.9 in Chap. 1. Then $\zeta + \zeta'$ is

$$0 \longrightarrow N \longrightarrow F \longrightarrow M \longrightarrow 0.$$

The set of equivalence classes $\mathscr{E}(M, N)$ of extensions of M by N together with the sum of extensions is an abelian group, and the class of the split extension is the zero element of that group.

There is an isomorphism of groups $\mathscr{E}(M, N) \to \mathrm{Ext}^1(M, N)$ which is defined as follows. Let $\zeta : 0 \to N \overset{u'}{\to} E \overset{v'}{\to} M \to 0$ be a representative of a class in $\mathscr{E}(M, N)$, and let $0 \to P_1 \overset{f}{\to} P_0 \overset{g}{\to} M \to 0$ be a projective resolution. Then since P_0 is projective, it follows that there exists a morphism $f' \in \mathrm{Hom}(P_0, E)$ such that $g = v' f'$. Now since ζ is exact, the universal property of the kernel implies that there exists also a morphism $u \in \mathrm{Hom}(P_1, N)$ such that the following diagram commutes:

$$\begin{array}{ccccccccc} & & P_1 & \xrightarrow{f} & P_0 & \xrightarrow{g} & M & \longrightarrow & 0 \\ & & \downarrow u & & \downarrow f' & & \downarrow = & & \\ \zeta : & 0 \longrightarrow & N & \xrightarrow{u'} & E & \xrightarrow{v'} & M & \longrightarrow & 0. \end{array}$$

Recall that $\mathrm{Ext}^1(M, N)$ is the cokernel of f^*, which is the quotient $\mathrm{Hom}(P_1, N)/\mathrm{im}\, f^*$. The isomorphism $\mathscr{E}(M, N) \to \mathrm{Ext}^1(M, N)$ is sending the class of ζ to the class of u.

Example 2.13. Let us compute the sum of the two short exact sequences ζ and ζ' in Example 2.12. Thus Q is the quiver

$$1 \underset{\beta}{\overset{\alpha}{\rightrightarrows}} 2\,,$$

$N = S(2)$, $M = S(1)$ and

$$E = k \underset{0}{\overset{1}{\rightrightarrows}} k \quad \text{and} \quad E' = k \underset{1}{\overset{0}{\rightrightarrows}} k\,.$$

Using our notation $E = (E_i, \varphi_\alpha)_{i \in Q_0, \alpha \in Q_1}$ and $E' = (E'_i, \varphi'_\alpha)_{i \in Q_0, \alpha \in Q_1}$, we have

$$\begin{array}{l} E_1 \cong E_2 \cong k \;\; \varphi_\alpha = 1 \;\; \varphi_\beta = 0 \\ E'_1 \cong E'_2 \cong k \;\; \varphi'_\alpha = 0 \;\; \varphi'_\beta = 1 \end{array}$$

Let us denote the elements of E as pairs $(e_1, e_2) \in E_1 \oplus E_2$ and those of E' as $(e'_1, e'_2) \in E'_1 \oplus E'_2$.

To compute the sum $\zeta + \zeta'$, we first need to compute the pull back E''. By definition

$$E'' = \left\{ \big((e_1, e_2), (e'_1, e'_2)\big) \in E \times E' \mid g(e_1, e_2) = g'(e'_1, e'_2) \right\}.$$

Since g and g' are both projections on the first component, we have

$$E'' = \left\{ \big((e_1, e_2), (e_1, e'_2)\big) \in E \times E' \right\}.$$

We want to write E'' as a representation $E'' = (E''_i, \varphi''_\alpha)_{i \in Q_0, \alpha \in Q_1}$. Our computation above shows that $E''_1 \cong k$ and $E''_2 \cong k^2$. Let's compute φ''_α and φ''_β. We have

$$\begin{array}{l} \varphi''_\alpha((e_1, e_2), (e_1, e'_2)) = (\varphi_\alpha(e_1, e_2), \varphi'_\alpha(e_1, e'_2)) = ((0, e_1), (0, 0)) \\ \varphi''_\beta((e_1, e_2), (e_1, e'_2)) = (\varphi_\beta(e_1, e_2), \varphi'_\beta(e_1, e'_2)) = ((0, 0), (0, e_1)) \end{array}$$

This shows that

$$E'' = k \underset{\begin{bmatrix} 0 \\ 1 \end{bmatrix}}{\overset{\begin{bmatrix} 1 \\ 0 \end{bmatrix}}{\rightrightarrows}} k^2.$$

Now we compute F. By definition and using the fact that both f and f' are inclusions in the second component, we have

$$F = E''/\{((0,n),(0,-n)) \mid n \in k\}.$$

Writing $F = (F_i, \psi_\alpha)$, we see that $F_1 \cong F_2 \cong k$. Moreover, if $\big((e_1,e_2),(e_1,e_2')\big) \in E''$ and $\overline{\big((e_1,e_2),(e_1,e_2')\big)}$ denotes its class in F, then

$$\psi_\alpha \overline{\big((e_1,e_2),(e_1,e_2')\big)} = \overline{\varphi''_\alpha\big((e_1,e_2),(e_1,e_2')\big)} = \overline{((0,e_1),(0,0))}$$

and

$$\psi_\beta \overline{\big((e_1,e_2),(e_1,e_2')\big)} = \overline{\varphi''_\beta\big((e_1,e_2),(e_1,e_2')\big)} = \overline{((0,0),(0,e_1))}.$$

In particular, $\psi_\alpha \overline{\big((e_1,e_2),(e_1,e_2')\big)} = -\psi_\beta \overline{\big((e_1,e_2),(e_1,e_2')\big)}$. This shows that

$$F \cong k \underset{-1}{\overset{1}{\rightrightarrows}} k.$$

Finally, we see that the sum $\zeta + \zeta'$ is the short exact sequence

$$0 \longrightarrow S(2) \xrightarrow{f''} F \xrightarrow{g''} S(1) \longrightarrow 0$$

with f'' the inclusion in the second component and g'' the projection on the first component.

Problems

Exercises for Chap. 2

2.1. Let Q be the quiver $1 \to 3 \to 4$, $1 \to 2 \to 3$. Prove that

$$P(1) \cong k \xrightarrow{\begin{bmatrix} x_1 \\ x_2 \end{bmatrix}} k^2 \xrightarrow{\begin{bmatrix} a & b \\ c & d \end{bmatrix}} k^2,$$

with $k \xrightarrow{z} k \xrightarrow{\begin{bmatrix} y_1 \\ y_2 \end{bmatrix}} k^2$ (from $P(1)$ at vertex 1 through vertex 2 to vertex 3) if and only if

1. all four maps have maximal rank, that is, $ad - bc \neq 0$, $(x_1, x_2) \neq (0, 0)$, $(y_1, y_2) \neq (0, 0)$, $z \neq 0$ and
2. the vectors $\begin{bmatrix} a & b \\ c & d \end{bmatrix}\begin{bmatrix} x_1 \\ x_2 \end{bmatrix}$ and $\begin{bmatrix} a & b \\ c & d \end{bmatrix}\begin{bmatrix} y_1 \\ y_2 \end{bmatrix}$ are linearly independent.

2.2. Compute the indecomposable projective representations $P(i)$ and the indecomposable injective representations $I(i)$ for the following quivers:

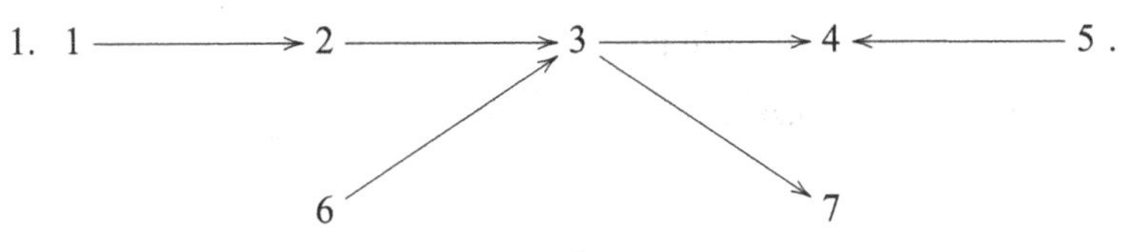

2. $1 \leftleftarrows 2 \leftleftarrows 3\,.$

3. $1 \rightrightarrows 2 \rightrightarrows 3\,.$ (with an arrow $1 \to 3$, and arrows $2 \to 4$, $3 \to 4$)

2.3. 1. Compute a projective resolution for each simple representation $S(i)$ for each of the quivers in (2.2).
2. Compute the dimension vector of $\tau S(i)$ for each simple representation $S(i)$ of the quivers in (2.2).

2.4. Prove Proposition 2.3

2.5. Prove Proposition 2.5

2.6. Show that for each $i \in Q_0$, the sequence

$$0 \longrightarrow \operatorname{rad} P(i) \longrightarrow P(i) \longrightarrow S(i) \longrightarrow 0$$

is a minimal projective resolution.

2.7. Let $M = (M_i, \varphi_\alpha)$ be a representation of Q. Prove that for any vertex i in Q, there is an isomorphism of vector spaces:

$$\operatorname{Hom}(M, I(i)) \cong M_i.$$

2.8. Let i and j be vertices in the quiver Q. Prove that the vector space $\operatorname{Hom}(I(i), I(j))$ has a basis consisting of all paths from j to i in Q. In particular, $\operatorname{End}(I(i)) \cong k$.

2.9. Prove that the representation $I(j)$ is a simple representation if and only if $\operatorname{Hom}(I(j), I(i)) = 0$ for all $i \neq j$.

2.10. Prove Proposition 2.27.

2.11. Prove that P is projective if and only if $\operatorname{Ext}^1(P, N) = 0$ for all representations N.

2.12. Let Q be the quiver

$$1 \underset{\beta}{\overset{\alpha}{\longleftleftarrows}} 2 \,,$$

and let M_λ be the representation defined in Exercise 1.6 of Chap. 1.

1. Show that the short exact sequences

$$0 \longrightarrow 1 \longrightarrow M_\lambda \longrightarrow 2 \longrightarrow 0,$$
$$0 \longrightarrow 1 \longrightarrow M_\mu \longrightarrow 2 \longrightarrow 0,$$

   are not equivalent if $\lambda \neq \mu$.
2. Show that $\tau M_\lambda = M_\lambda$, for all λ.

Chapter 3
Examples of Auslander–Reiten Quivers

We have already pointed out in Sect. 1.5 that Auslander–Reiten quivers provide a threefold information about the representation theory of the quiver, namely the indecomposable representations, the irreducible morphisms, and the almost split sequences—these in turn should be thought of the building blocks of arbitrary representations, morphisms, and short exact sequences, respectively.

We have developed enough of the theory by now to be able to compute and appreciate Auslander–Reiten quivers. We present here several different methods of computation, although we are not able yet to prove that these methods actually produce the desired result; this justification is postponed to Chap. 7.

This chapter is subdivided into several sections. In the first section, we compute Auslander–Reiten quivers of type $\mathbb{A}_n$, the second section is a digression on finite representation type, and the third section treats the Auslander–Reiten quivers of type $\mathbb{D}_n$. In both the first and the third section, we present several methods to compute the Auslander–Reiten quiver.

The first method, the *knitting algorithm*, is a recursive procedure which owes its name to the fact that it produces one mesh after the other. The second method is to compute the orbits under the Auslander–Reiten translation τ. While the knitting algorithm produces the Auslander–Reiten quiver by computing the next vertical cross section and gradually progressing from left to right, the τ-orbit procedure computes horizontal cross sections of the Auslander–Reiten quiver. The third method is a geometric construction of the Auslander–Reiten quiver in terms of diagonals in a polygon in type $\mathbb{A}_n$ and in terms of arcs in a punctured polygon in type $\mathbb{D}_n$. We then show how to use the Auslander–Reiten quiver to compute the dimensions of Hom and Ext spaces between modules. In the fourth section, we introduce bound quivers and their representations in order to show how the geometric constructions for type $\mathbb{A}_n$ and $\mathbb{D}_n$ naturally generalize to the so-called cluster-tilted bound quivers. The reader who is not enthusiastic about the geometric realizations may very well skip the subsection on the punctured polygon (Sect. 3.3.3).

R. Schiffler, *Quiver Representations*, CMS Books in Mathematics,
DOI 10.1007/978-3-319-09204-1_3

3.1 Auslander–Reiten Quivers of Type $\mathbb{A}_n$

In this section, let Q be a quiver of type $\mathbb{A}_n$, that is, the underlying unoriented graph of Q is the Dynkin diagram of type $\mathbb{A}_n$:

$$1 \text{ --- } 2 \text{ --- } 3 \text{ --- } \dots \text{ --- } (n-1) \text{ --- } n.$$

We will see several ways to construct the Auslander–Reiten quiver of Q.

3.1.1 The Knitting Algorithm

The knitting algorithm owes its name to the fact that it recursively constructs one mesh after the other, from left to right. In order to get started one has to compute the indecomposable projective representations which are the leftmost indecomposable representations in the Auslander–Reiten quiver.

1. Compute the indecomposable projective representations

$$P(1), P(2), \dots, P(n).$$

2. Draw an arrow $P(i) \to P(j)$ whenever there exists an arrow $j \to i$ in Q_1, in such a way that each $P(i)$ sits at a different level.
3. (Knitting) There are three types of meshes. Complete each mesh as shown in Fig. 3.1 in such a way that

$$\underline{\dim}\, L + \underline{\dim}\, \tau^{-1} L = \sum_{i=1}^{2} \underline{\dim}\, M_i.$$

4. Repeat step 3 until you get negative integers in the dimension vector.

Observe that, every time we perform the third step, the representations L and M_i have been computed earlier and only $\tau^{-1}L$ is unknown.

The isoclasses of indecomposable representations of quivers of type $\mathbb{A}_n$ are determined by their dimension vectors as follows. The dimension vector is always of the form $(0, \dots, 0, 1, \dots, 1, 0, \dots, 0)$, and the corresponding representation is $M = (M_i, \varphi_\alpha)$ with $M_i = k$ if the dimension at i is one, and $M_i = 0$ otherwise; and $\varphi_\alpha = 1$ if the dimension at $s(\alpha)$ and at $t(\alpha)$ is one, and $\varphi_\alpha = 0$ otherwise.

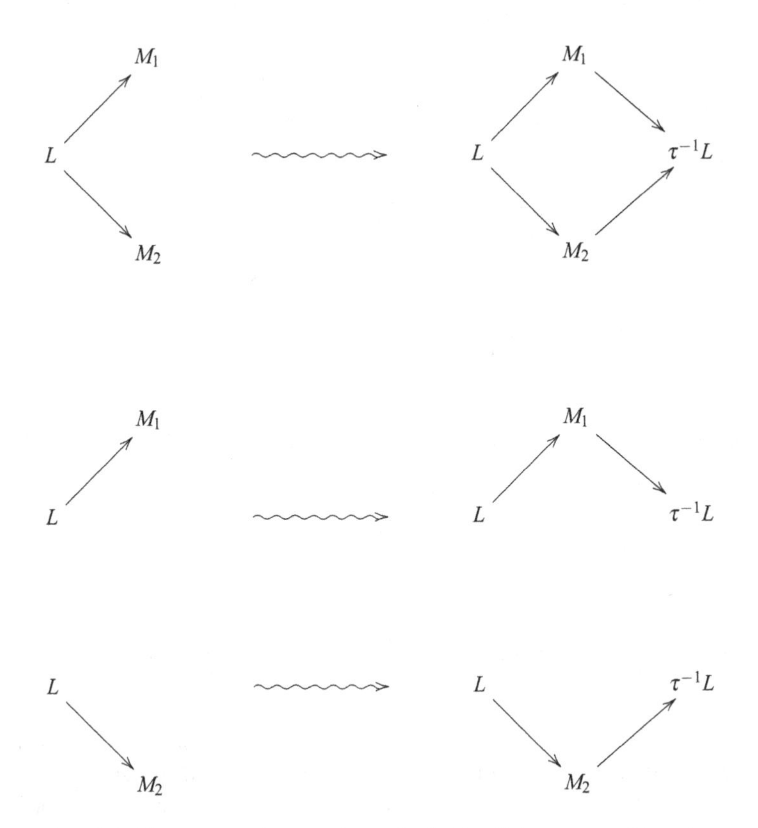

Fig. 3.1 Three types of meshes in the Auslander–Reiten quiver of type $\mathbb{A}_n$

Example 3.1. Let Q be the quiver

$$1 \longleftarrow 2 \longleftarrow 3 \longrightarrow 4 \longleftarrow 5\,.$$

Then

$$P(1) = 1 \;\; P(2) = \begin{smallmatrix} 2 \\ 1 \end{smallmatrix} \;\; P(3) = \begin{smallmatrix} & 3 & \\ 2 & & 4 \\ 1 & & \end{smallmatrix}$$

$$P(4) = 4 \;\; P(5) = \begin{smallmatrix} 5 \\ 4 \end{smallmatrix}$$

and the Auslander–Reiten quiver is

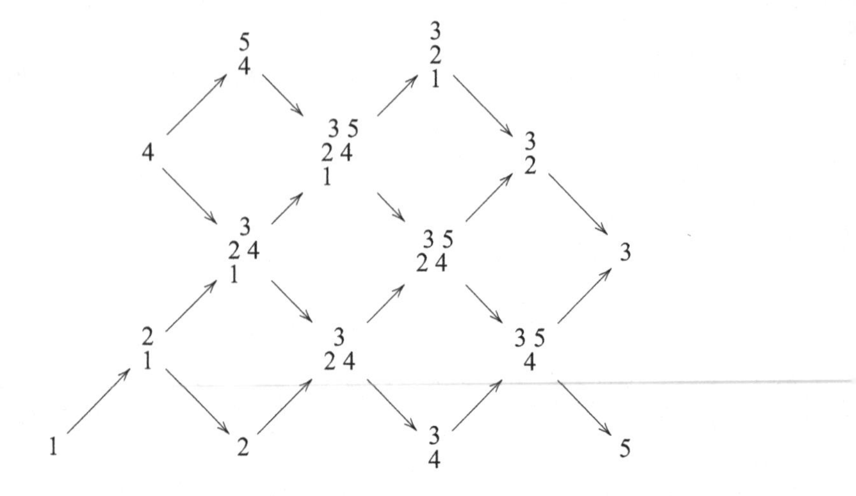

3.1.2 τ-Orbits

The map τ is the Auslander–Reiten translation. In the Auslander–Reiten quiver, it is the translation that sends the rightmost point of a mesh to the leftmost point of the same mesh. The τ-orbit of an indecomposable representation is the set of all representations that can be obtained by applying τ or τ^{-1} repeatedly to the representation. Thus the τ-orbits in the Auslander–Reiten quivers of type $\mathbb{A}_n$ consist of the representations that sit on the same level in the quiver.

Each τ-orbit in the Auslander–Reiten quiver of type $\mathbb{A}_n$ contains exactly one projective representation, so starting from the projectives, we can compute the whole quiver by computing the τ-orbits.

There are several methods to compute τ-orbits.

3.1.2.1 First Method: Auslander–Reiten Translation

Let M be an indecomposable representation that is not injective. We want to compute the translation to the right $\tau^{-1}M$ of M. Start with an injective resolution

$$0 \longrightarrow M \longrightarrow I_0 \xrightarrow{g} I_1 \longrightarrow 0,$$

and apply the inverse Nakayama functor ν^{-1}. This functor maps the indecomposable injective representation $I(j)$ to the corresponding indecomposable projective representation $P(j)$; see Proposition 2.29 of Chap. 2. Then $\tau^{-1}M$ is given by the projective resolution:

$$0 \longrightarrow \nu^{-1}I_0 \xrightarrow{\nu^{-1}(g)} \nu^{-1}I_1 \longrightarrow \tau^{-1}M \longrightarrow 0.$$

Let us compute $\tau^{-1}M$ for the module $M = 4$ in Example 3.1. The upper line in the following diagram shows an injective resolution of M, and the lower line shows the corresponding projective resolution of $\tau^{-1}M$ obtained by applying ν^{-1}:

$$\begin{array}{ccccccccccc} 0 & \longrightarrow & 4 & \longrightarrow & \begin{smallmatrix} 3\,5 \\ 4 \end{smallmatrix} & \longrightarrow & 3 \oplus 5 & \longrightarrow & 0 & & \\ & & & & \downarrow{\scriptstyle \nu^{-1}} & & \downarrow{\scriptstyle \nu^{-1}} & & & & \\ & & 0 & \longrightarrow & 4 & \longrightarrow & \begin{smallmatrix} 3 \\ 2\,4 \\ 1 \end{smallmatrix} \oplus \begin{smallmatrix} 5 \\ 4 \end{smallmatrix} & \longrightarrow & \begin{smallmatrix} 3\,5 \\ 2\,4 \\ 1 \end{smallmatrix} & \longrightarrow & 0 \end{array}$$

Thus $\tau^{-1}M = \begin{smallmatrix} 3\,5 \\ 2\,4 \\ 1 \end{smallmatrix}$ which verifies the result of Example 3.1.

3.1.2.2 Second Method: Coxeter Functor

Choose a sequence of vertices $(i_1, i_2, \ldots, i_n)$, with $i_j \neq i_\ell$ if $i \neq \ell$, as follows:

i_1 is a sink of Q;
i_2 is a sink of the quiver $s_{i_1} Q$ obtained from Q by reversing all arrows that are incident to the vertex i_1;
i_t is a sink of $s_{i_{t-1}} \ldots s_{i_2} s_{i_1} Q$, for $t = 2, 3, \ldots, n$.

Thus in Example 3.1 such a sequence would be $(1, 4, 2, 3, 5)$.

Next, we need the notion of reflections $s_i : \mathbb{R}^n \to \mathbb{R}^n$ defined by $s_i(x) = x - 2B(x, e_i)e_i$, where $\{e_1, \ldots, e_n\}$ is a basis of $\mathbb{R}^n$ and B is a symmetric bilinear form defined by

$$B(e_i, e_j) = \begin{cases} 1 & \text{if } i = j \\ -1/2 & \text{if } i \text{ is adjacent to } j \text{ in } Q \\ 0 & \text{otherwise.} \end{cases}$$

In other words, $s_i(\sum_j a_j e_j) = \sum_j a'_j e_j$, where $a'_j = a_j$ if $j \neq i$ and $a'_i = -a_i + \sum_{i \,—\, j} a_j$, where the sum is over all vertices j that are adjacent to i in Q.

Finally, we define a so-called *Coxeter element* $c = s_{i_1} s_{i_2} \cdots s_{i_n}$ as a product of reflections using the sequence of vertices defined above. Thus in Example 3.1 such a Coxeter element would be $c = s_1 s_4 s_2 s_3 s_5$.

One can use this Coxeter element to compute the dimension vector of the representation $\tau^{-1}M$ from the dimension vector of M. If $\underline{\dim}M = (d_1, d_2, \ldots, d_n)$, then $c(\sum_i d_i e_i) = \sum_i d_i' e_i$ and $\underline{\dim}(\tau^{-1}M) = (d_1', d_2', \ldots, d_n')$.

Let us use this method to compute the dimension vector of $\tau^{-1}4$ in Example 3.1. We have $\underline{\dim}\, M = (0, 0, 0, 1, 0)$. Thus $\underline{\dim}\, \tau^{-1}M$ is equal to

$$\begin{aligned} s_1 s_4 s_2 s_3 s_5(e_4) &= s_1 s_4 s_2 s_3(e_4 + e_5) \\ &= s_1 s_4 s_2(e_3 + e_4 + e_5) \\ &= s_1 s_4(e_2 + e_3 + e_4 + e_5) \\ &= s_1(e_2 + e_3 + e_4 + e_5) \\ &= e_1 + e_2 + e_3 + e_4 + e_5, \end{aligned}$$

which again confirms the result obtained in Example 3.1.

Another way of defining the action of the Coxeter element is to use the **Cartan matrix** C of the quiver Q. This matrix is defined as $C = (c_{ij})_{1 \le i,j \le n}$, where c_{ij} is the number of paths from j to i and n is the number of vertices in Q. It follows directly from the definition that, for every vertex i, the ith column of C is exactly the dimension vector of the indecomposable projective representation $P(i)$ and the ith row of C is exactly the dimension vector of the indecomposable injective representation $I(i)$.

Since Q has no oriented cycles, we can always renumber the vertices of Q in such a way that, if there is a path from j to i, then $i \le j$; in other words, there is a renumbering of the vertices such that the matrix C is upper triangular. Also note that the diagonal entries of C are all equal to 1, since there is exactly one path, the constant path, from each vertex to itself. This shows that C is invertible.

Its inverse C^{-1} is the matrix $(b_{ij})_{1 \le i,j \le n}$ where $b_{ii} = 1$, and if $i \ne j$, then $-b_{ij}$ is the number of arrows from j to i in Q. To show that this is indeed the inverse of C, we multiply the two matrices:

$$(c_{ij})_{i,j}(b_{j\ell})_{j,\ell} = \left(\sum_j c_{ij} b_{j\ell}\right)_{i,\ell}.$$

Note first that the diagonal entries $\sum_j c_{ij} b_{ji} = c_{ii} b_{ii} = 1$, since both matrices are upper triangular (up to some renumbering of the vertices). Next, if $i \ne \ell$, then each path from ℓ to i must start with some arrow from ℓ to some vertex j. Therefore, the number $c_{i\ell}$ of paths from ℓ to i can be computed as

$$c_{i\ell} = \sum_{j \in Q_0 \setminus \{\ell\}} c_{ij}(-b_{j\ell}).$$

Now using $b_{\ell\ell} = 1$, we have $\sum_j c_{ij} b_{j\ell} = c_{i\ell} + \sum_{j \ne \ell} c_{ij} b_{j\ell} = 0$ if $i \ne \ell$. Thus $C^{-1} = (b_{ij})_{1 \le i,j \le n}$.

Now we define yet another matrix, the **Coxeter matrix** Φ, as $\Phi = -C^{\top}(C^{-1})$, and its inverse is $\Phi^{-1} = -C(C^{-1})^{\top}$, the superscript $^{\top}$ here denotes the transpose of a matrix. Then

$$\Phi\,\underline{\dim}\,M = \underline{\dim}\,\tau M, \text{ if } M \text{ is not projective and } \Phi\,\underline{\dim}\,P(j) = -\underline{\dim}\,I(j),$$

whereas

$$\Phi^{-1}\underline{\dim}\,M = \underline{\dim}\,\tau^{-1}M, \text{ if } M \text{ is not injective and } \Phi^{-1}\underline{\dim}\,I(j) = -\underline{\dim}\,P(j).$$

In our Example 3.1, we have

$$C = \begin{bmatrix} 1&1&1&0&0\\ 0&1&1&0&0\\ 0&0&1&0&0\\ 0&0&1&1&1\\ 0&0&0&0&1 \end{bmatrix} \qquad (C^{-1}) = \begin{bmatrix} 1&-1&0&0&0\\ 0&1&-1&0&0\\ 0&0&1&0&0\\ 0&0&-1&1&-1\\ 0&0&0&0&1 \end{bmatrix}$$

$$\Phi = \begin{bmatrix} -1&1&0&0&0\\ -1&0&1&0&0\\ -1&0&1&-1&1\\ 0&0&1&-1&1\\ 0&0&1&-1&0 \end{bmatrix} \qquad \Phi^{-1} = \begin{bmatrix} 0&0&-1&1&0\\ 1&0&-1&1&0\\ 0&1&-1&1&0\\ 0&1&-1&1&-1\\ 0&0&0&1&-1 \end{bmatrix}$$

so that the dimension of $\tau^{-1}4$ can be computed by $\Phi^{-1}\,(0,0,0,1,0)^{\top}$ which is equal to $(1,1,1,1,1)^{\top}$. On the other hand, $\Phi\,\underline{\dim}\,P(4) = \Phi\,(0,0,0,1,0)^{\top} = (0,0,-1,-1,-1)^{\top} = -\underline{\dim}\,I(4)$.

3.1.3 Diagonals of a Polygon with n + 3 Vertices

In this section, we give a geometric way to construct the Auslander–Reiten quiver of a quiver Q of type $\mathbb{A}_n$ from a triangulation of a polygon. This method works only for quivers of type $\mathbb{A}_n$.

Start with a regular polygon with $n+3$ vertices. A *diagonal* in the polygon is a straight line segment that joins two of the vertices and goes through the interior of the polygon, and a *triangulation* of the polygon is a maximal set of non-crossing diagonals. Such a triangulation cuts the polygon into triangles, hence the name. Given a triangle with sides a, b, c, we say that the side a is clockwise of the side b if going along the boundary of the triangle in the clockwise direction corresponds to the sequence $a, b, c, a, b, c, a \ldots$.

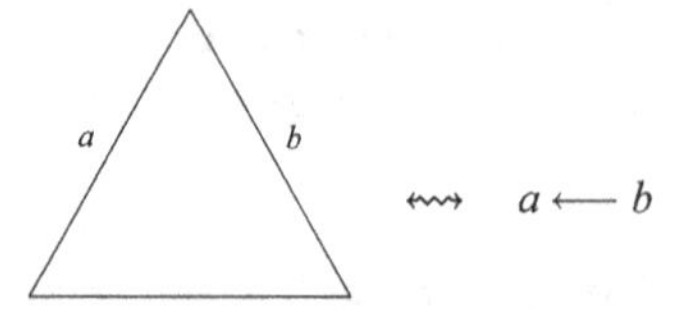

Fig. 3.2 b is clockwise from a corresponding to an arrow from b to a

We will associate a triangulation T_Q to our type $\mathbb{A}_n$ quiver Q as follows: Let 1 be a vertex in the quiver that has only one neighbor. Draw a diagonal that cuts off a triangle Δ_0 and label that diagonal 1. If $1 \leftarrow 2$ is an arrow in Q, then draw the unique diagonal 2 such that 1, 2 and one boundary segment of the polygon form a triangle Δ_1 in such a way that diagonal 2 is clockwise of diagonal 1 in the triangle Δ_1. If, on the other hand, $1 \rightarrow 2$ is an arrow in Q, draw the unique diagonal 2 such that diagonal 2 is counterclockwise of diagonal 1 in the triangle Δ_1; see Fig. 3.2. Continue this procedure up to diagonal n.

In this way the quiver

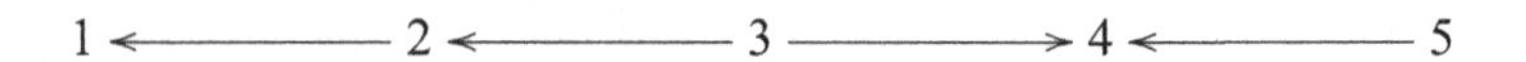

of Example 3.1 gives rise to the triangulation

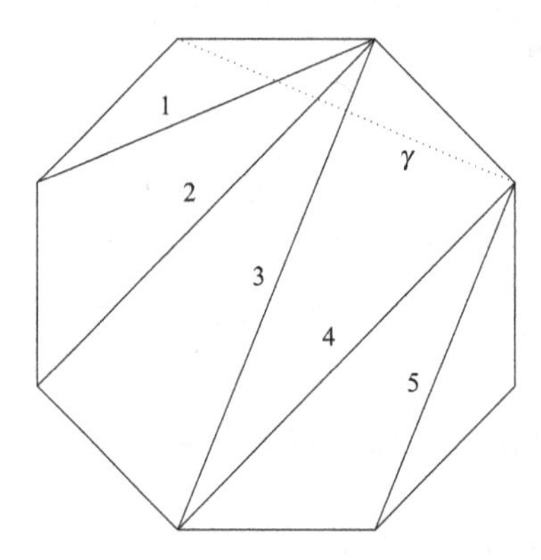

Since T_Q is a triangulation of the polygon, any other diagonal γ which is not already in T_Q will cut through a certain number of diagonals in T_Q; in fact, any such diagonal γ is uniquely determined by the set of diagonals in T_Q that γ crosses. To such a diagonal γ, we associate a representation $M_\gamma = (M_i, \varphi_\alpha)$ of Q by letting

$$M_i = \begin{cases} \text{k if the diagonal } \gamma \text{ crosses the diagonal } i; \\ 0 \text{ otherwise;} \end{cases}$$

and setting $\varphi_\alpha = 1$ whenever $M_{s(\alpha)} = M_{t(\alpha)} = k$, and $\varphi_\alpha = 0$ otherwise. In the example, the diagonal γ crosses the diagonals 1, 2, and 3, and the corresponding representation is

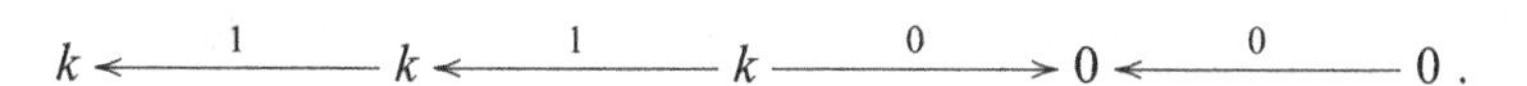

The map $\gamma \mapsto M_\gamma$ is a bijection from the set of diagonals that are not in T_Q and the set of isoclasses of indecomposable representations of Q.

The Auslander–Reiten translation τ is given by an elementary clockwise rotation of the polygon, so in our example τ of γ is the diagonal that cuts through the diagonals 4 and 5.

The projective representation $P(i)$ is given by τ^{-1} of the diagonal i, and the injective representation $I(i)$ is given by τ of the diagonal i. In our example $P(1)$ is the diagonal that cuts through the diagonal 1 only and $I(1)$ is the diagonal γ.

The complete Auslander–Reiten quiver can be easily constructed now starting with the projectives and applying the elementary rotation to compute the τ-orbits until we reach the injective in each τ-orbit, and the Auslander–Reiten quiver is

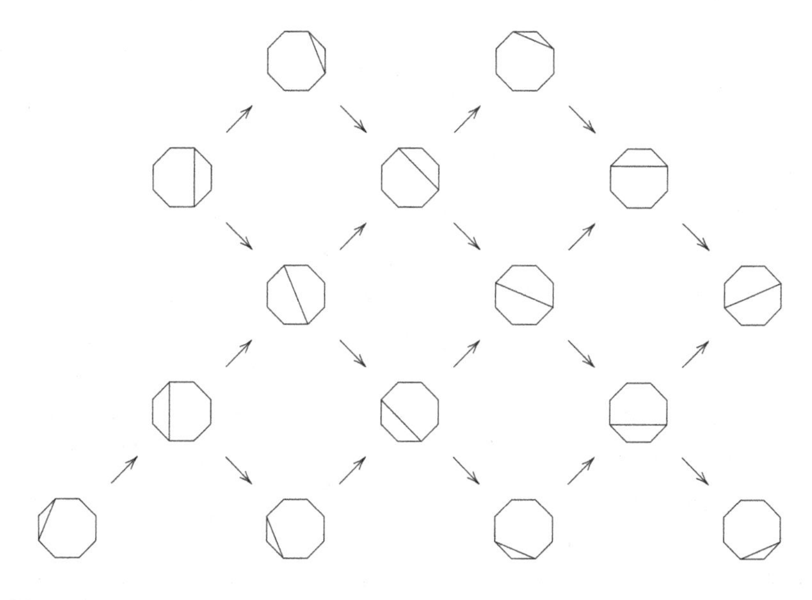

Note that any arrow in the Auslander–Reiten quiver acts on the diagonal by pivoting one of the endpoints of the diagonal to its counterclockwise neighbor.

3.1.4 Computing Hom Dimensions, Ext Dimensions, and Short Exact Sequences

Given two indecomposable representations M and N, we want to have information about the space of morphisms $\mathrm{Hom}(M, N)$. The Auslander–Reiten quiver allows us to compute the dimension of this space easily, at least if M and N lie in the same connected component.

3.1.4.1 Dimension of Hom(M, N)

Let Q be a type $\mathbb{A}$ quiver and let M, N be two indecomposable representations of Q. We can compute the dimension of the vector space $\mathrm{Hom}(M, N)$ using the relative position of M and N in the Auslander–Reiten quiver. For this we need to introduce some terminology:

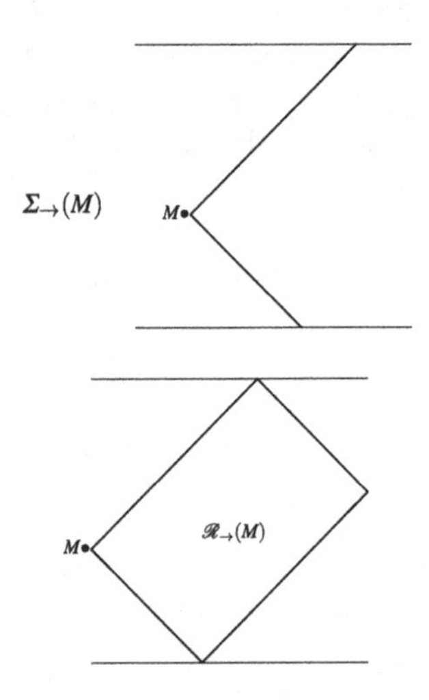

A path $M_0 \to M_1 \to \cdots \to M_s$ in the Auslander–Reiten quiver is called a **sectional path** if $\tau M_{i+1} \neq M_{i-1}$ for all $i = 1, \ldots, s-1$. Let $\Sigma_{\to}(M)$ be the set of all indecomposable representations that can be reached from M by a sectional path, and let $\Sigma_{\leftarrow}(M)$ be the set of all indecomposable representations from which one can reach M by a sectional path.

Now let $\mathscr{R}_{\to}(M)$ be the set of all indecomposable representations whose position in the Auslander–Reiten quiver is in the slanted rectangular region whose left boundary is $\Sigma_{\to}(M)$. We call $\mathscr{R}_{\to}(M)$ the maximal slanted rectangle in the Auslander–Reiten quiver whose leftmost point is M. Then $\dim \mathrm{Hom}(M, N)$ is either 1 or 0, and it is 1 if and only if N lies in $\mathscr{R}_{\to}(M)$.

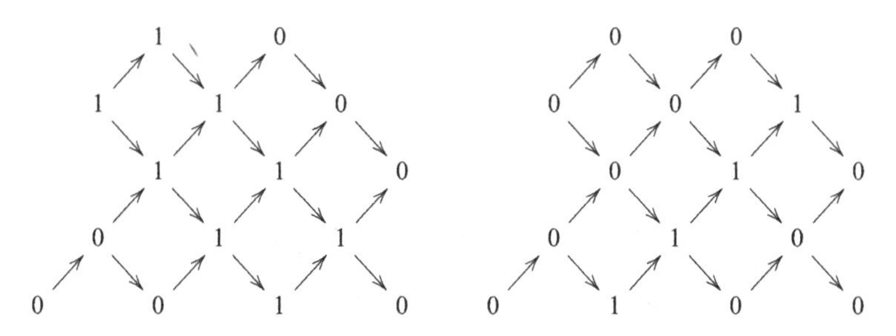

Fig. 3.3 Dimension of $\mathrm{Hom}(M, -)$ for $M = P(4)$ on the *left* and $M = S(2)$ on the *right*. The position of the representation M is at the leftmost 1 in each case; the numbers 0, 1 indicate the dimension of $\mathrm{Hom}(M, N)$ for each indecomposable representation N

We illustrate this concept in Fig. 3.3 for the Auslander–Reiten quiver of Example 3.1. On the left side of Fig. 3.3, the module M is the indecomposable projective $P(4)$. Its position in the Auslander–Reiten quiver is the leftmost 1 in the figure, so this 1 indicates that $\dim \mathrm{Hom}(M, M) = 1$. A basis for this vector space is the identity morphism 1_M. Each indecomposable representation N is located at a specific point in the Auslander–Reiten quiver; the number 0 or 1 at that point indicates the dimension of $\mathrm{Hom}(M, N)$ for each N.

In the Auslander–Reiten quiver on the right-hand side of Fig. 3.3, the module M is the simple module $S(2)$. Again its position is the leftmost 1 in that figure. The rectangle on which $\mathrm{Hom}(M, -)$ is nonzero reduces in this case to a single line.

Symmetrically, we denote by $\mathscr{R}_{\leftarrow}(N)$ the maximal slanted rectangle in the Auslander–Reiten quiver whose rightmost point is N. We can compute the dimension of $\mathrm{Hom}(-, N)$ using $\mathscr{R}_{\leftarrow}(N)$. Thus the data in the left picture in Fig. 3.3 also computes the $\dim \mathrm{Hom}(-, N)$ for $N = I(4)$.

Note that if $M = P(i)$ is an indecomposable projective, then it follows from Theorem 2.11 that the representations in $\mathscr{R}_{\rightarrow}(P(i))$ are precisely the indecomposable representations N such that $N_i \neq 0$. It then follows from Exercise 2.7 of Chap. 2 that there is a unique rightmost point in $\mathscr{R}_{\rightarrow}(P(i))$ which must be the position of the indecomposable injective representation $I(i)$. In particular, $\mathscr{R}_{\rightarrow}(P(i)) = \mathscr{R}_{\leftarrow}(I(i))$.

Figure 3.4 shows an example where the right end of the $\mathscr{R}_{\rightarrow}(M)$ does not really have the shape of a rectangle, because the Auslander–Reiten quiver ends before the rectangle is completed. This happens exactly when M is not projective.

3.1.4.2 Dimension of $\mathrm{Ext}^1(M, N)$

Next we compute the dimensions of the vector spaces $\mathrm{Ext}^1(M, N)$ for indecomposable representations M, N of type $\mathbb{A}$. If M is projective, then this space is zero, by Exercise 2.11 of Chap. 2, so let us assume that M is not projective. Thus τM is a

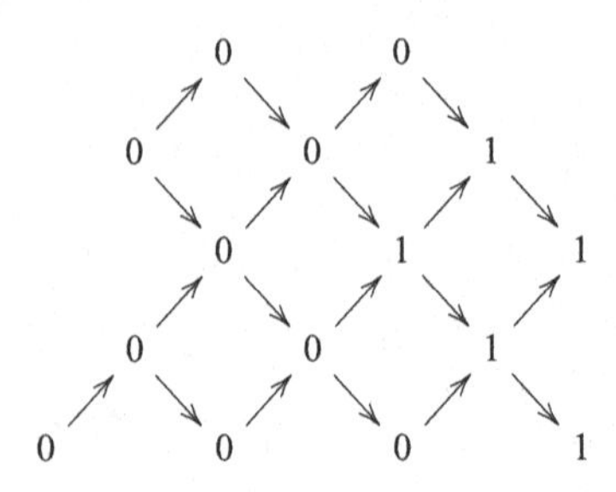

Fig. 3.4 Dimension of $\mathrm{Hom}(M,-)$ where M is the representation whose dimension vector is $(0,1,1,1,1)$

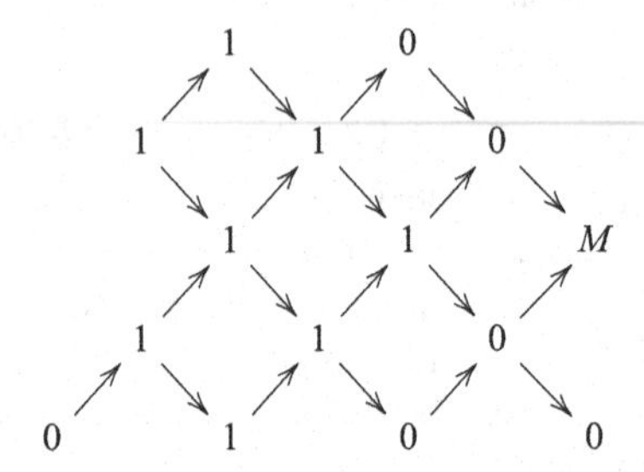

Fig. 3.5 Dimension of $\mathrm{Ext}^1(M,-)$ for $M = I(3)$

point in the Auslander–Reiten quiver. We will see in Theorem 7.18 that there is an isomorphism

$$\mathrm{Ext}^1(M,N) \cong D\mathrm{Hom}(N,\tau M),$$

where D is the duality and τ is the Auslander–Reiten translation. This isomorphism implies that $\dim \mathrm{Ext}^1(M,N) = \dim \mathrm{Hom}(N,\tau M)$ and therefore we can compute the dimension of $\mathrm{Ext}^1(M,-)$ using the maximal slanted rectangle $\mathscr{R}_{\leftarrow}(\tau M)$.

Figure 3.5 shows the dimension of $\mathrm{Ext}^1(M,-)$ for the representation $M = I(3)$ in our running example.

3.1.4.3 Short Exact Sequences

We have seen in Sect. 2.4 of Chap. 2 that the elements of $\mathrm{Ext}^1(M,N)$ can be represented by short exact sequences of the form $0 \to N \to E \to M \to 0$, where E is some representation of Q. We are interested here in the case where M and N are indecomposable—this does *not* imply that E is indecomposable.

We now want to compute the possible representations E for these short exact sequences. If the dimension of $\mathrm{Ext}^1(M,N)$ is 0, then the only possibility is $E \cong M \oplus N$. If on the other hand, the dimension of $\mathrm{Ext}^1(M,N)$ is 1, then, up to

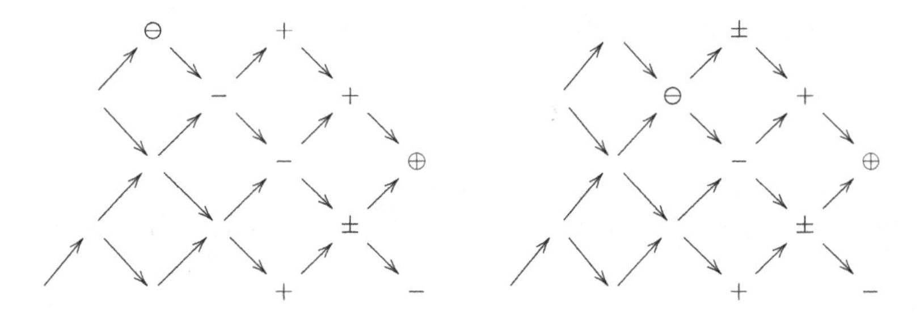

Fig. 3.6 Computing short exact sequences

isomorphism, there is exactly one other possibility for E. For representations of type $\mathbb{A}$, we can compute E simply from the relative positions of M and N in the Auslander–Reiten quiver:

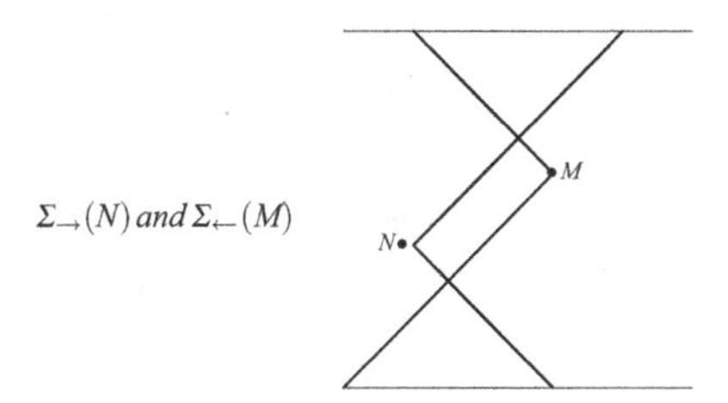

Let M, N be indecomposable representations of a quiver of type $\mathbb{A}$ such that $\mathrm{Ext}^1(M, N) \neq 0$. Then N must lie in $\mathscr{R}_{\leftarrow}(\tau M)$ and this implies that $\Sigma_{\rightarrow}(N)$ and $\Sigma_{\leftarrow}(M)$ have either 1 or 2 points in common, and these points correspond to the indecomposable summands of E.

We illustrate this situation in Fig. 3.6; the representation N is marked by $\ominus$ and the representation M by $\oplus$. The representations in $\Sigma_{\rightarrow}(N)$ are marked by $-$ or $\ominus$ (for N) and those in $\Sigma_{\leftarrow}(M)$ by $+$ or $\oplus$ (for M). The points of intersection are marked $\pm$. The example of the left-hand side of Fig. 3.6 corresponds to the short exact sequence:

$$0 \longrightarrow \begin{smallmatrix} 5 \\ 4 \end{smallmatrix} \longrightarrow \begin{smallmatrix} 3\,5 \\ 4 \end{smallmatrix} \longrightarrow 3 \longrightarrow 0$$

and the example on the right-hand side corresponds to the short exact sequence:

$$0 \longrightarrow \begin{smallmatrix} 3\,5 \\ 2\,4 \\ 1 \end{smallmatrix} \longrightarrow \begin{smallmatrix} 3\,5 \\ 4 \end{smallmatrix} \oplus \begin{smallmatrix} 3 \\ 2 \\ 1 \end{smallmatrix} \longrightarrow 3 \longrightarrow 0\,.$$

3.2 Representation Type

3.2.1 *Gabriel's Theorem: Finite Representation Type*

A quiver Q is said to be of **finite representation type** if the number of isoclasses of indecomposable representations of Q is finite. In this section, we list the quivers of finite representation type. It turns out that this classification depends only on the shape of the quiver and not on the particular orientation of the arrows. We therefore define the **underlying graph** of the quiver Q to be the graph obtained from Q by forgetting the direction of the arrows; thus the underlying graph has the same vertices as Q and for each arrow $i \to j$ in Q there is an edge $i — j$ in the underlying graph.

The graphs in Fig. 3.7 are called **Dynkin diagrams**. These graphs play an important role in mathematics when it comes to classifications. There are four infinite series, types $\mathbb{A}, \mathbb{B}, \mathbb{C}$ and $\mathbb{D}$, and five exceptional diagrams, types $\mathbb{E}, \mathbb{F}$ and $\mathbb{G}$. The types $\mathbb{A}, \mathbb{D}, \mathbb{E}$ are the only ones that have no parallel edges; these types are called **simply laced** Dynkin diagrams and will be of particular interest to us. The classification result is as follows:

Theorem 3.1 (Gabriel's Theorem, Part I). *A connected quiver is of finite representation type if and only if its underlying graph is one of the Dynkin diagrams of type* $\mathbb{A}, \mathbb{D}$ *or* $\mathbb{E}$.

This is a very surprising result, one might of course ask now what is so special about the Dynkin diagrams, or why are there only three diagrams of type $\mathbb{E}$? Note that we cannot come up with a diagram of $\mathbb{E}$ type with five or less vertices, because it would be a diagram of type $\mathbb{D}$ or $\mathbb{A}$. But what about $\mathbb{E}$ type diagrams with 9, 10, or more vertices? Well, the simple answer is that you then get infinitely many indecomposable representations, but this answer does not really settle the question: why?

One thing we can say is that we are not the only ones who are puzzled about this fact, because the Dynkin diagrams show up in finite type classifications of objects in several different fields of mathematics, for example, in the classifications of Lie algebras, root systems, Coxeter groups, and cluster algebras. These diagrams just happen to be very fundamental objects that reflect finite type structures that arise in nature.

We postpone the proof of Gabriel's theorem to Chap. 8. For now we just want to use it to move beyond type $\mathbb{A}$ in our section on examples of Auslander–Reiten quivers. From Gabriel's theorem we see that we should compute the $\mathbb{D}$-type next. This is done in the following section.

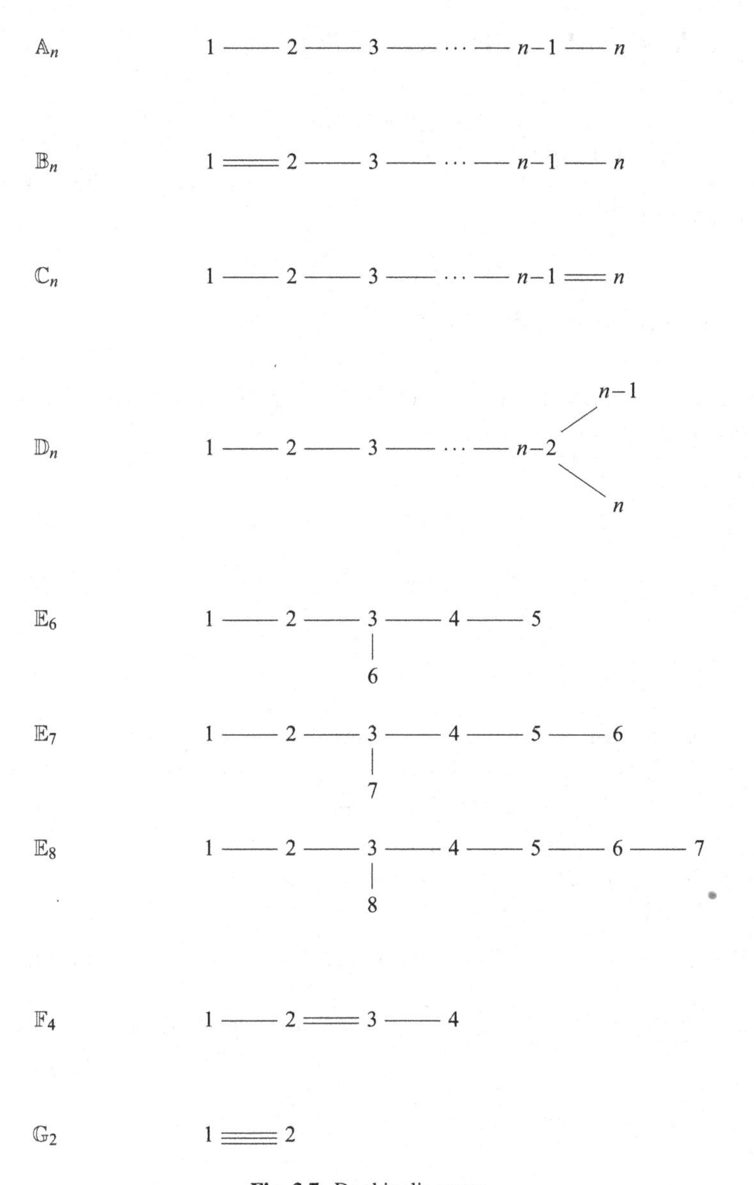

Fig. 3.7 Dynkin diagrams

3.3 Auslander–Reiten Quivers of Type $\mathbb{D}_n$

In this section, let Q be a quiver of type $\mathbb{D}_n$, that is, the underlying unoriented graph of Q is the Dynkin diagram of type $\mathbb{D}_n$.

We will use the different techniques from Sect. 3.1 to construct the Auslander–Reiten quiver of Q.

3.3.1 *The Knitting Algorithm*

We can use this algorithm in almost the same way as for type $\mathbb{A}_n$, with the difference that now, there is a fourth type of mesh:

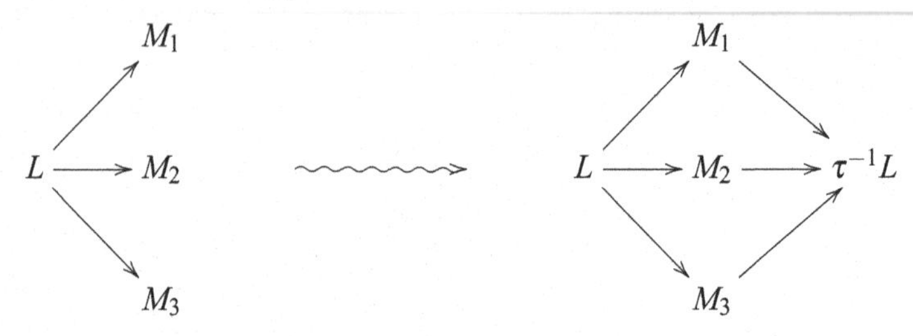

The isoclasses of indecomposable representations of quivers of type $\mathbb{D}_n$ are determined by their dimension vectors $\mathbf{d} = (d_1, \ldots, d_n)$ as follows. The entries d_i of the dimension vector are either 0, 1 or 2, and if we have $d_i = 2$, then

1. i is one of the vertices $2, 3, \ldots, n-2$,
2. for all vertices j with $i \leq j \leq n-2$ we have $d_j = 2$,
3. $d_{i-1} \geq 1$ and $d_{n-1} = d_n = 1$.

Thus the vertices i with $d_i = 2$ form a subgraph of type $\mathbb{A}$ that contains the vertex $n-2$.

The vertices $i \neq n-1, n$ with $d_i = 1$ also form a subgraph of type $\mathbb{A}$, and if $d_j \neq 2$ for all j, then all the vertices i with $d_i = 1$ form a subgraph of type $\mathbb{A}$ or a subgraph of type $\mathbb{D}$.

Graphically, we can represent some of these configurations as follows:

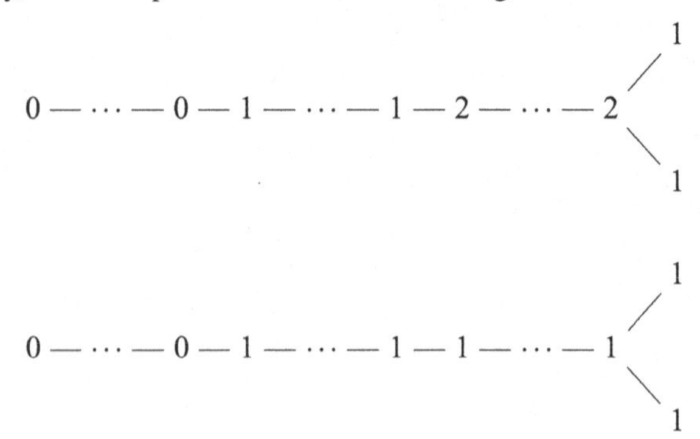

The corresponding representation is $M = (M_i, \varphi_\alpha)$ with $M_i = k^{d_i}$; and $\varphi_\alpha = 1$ if $d_{s(\alpha)} = d_{t(\alpha)}$, $\varphi_\alpha = 0$ if one of $d_{s(\alpha)}, d_{t(\alpha)}$ is zero. If one of the d_i is 2, then there are exactly three arrows that connect a vertex with dimension 1 to a vertex with dimension 2: two of these arrows, let us call them β_1, β_2, connect the vertex $n-2$ with the vertices $n-1$ and n, the vector space of dimension two being at $n-2$, while the third arrow α_i connects two vertices i and $i+1$, the vector space of dimension two being at vertex $i+1$. Consider the one-dimensional subspace of M_{i+1} given by

$$\begin{cases} \operatorname{im} \varphi_{\alpha_i} & \text{if } \alpha_i \text{ points to } i+1, \\ \ker \varphi_{\alpha_i} & \text{otherwise.} \end{cases}$$

Under the composition of the identity maps $\varphi_{\alpha_{n-3}} \cdots \varphi_{\alpha_{i+1}}$ this one-dimensional subspace is sent to a one-dimensional subspace ℓ_1 of M_{n-2}. Consider also the following two one-dimensional subspaces ℓ_2 and ℓ_3 of M_{n-2}:

$$\ell_2 = \begin{cases} \operatorname{im} \varphi_{\beta_1} & \text{if } \beta_1 \text{ points to } n-2, \\ \ker \varphi_{\beta_1} & \text{otherwise;} \end{cases}$$

and

$$\ell_3 = \begin{cases} \operatorname{im} \varphi_{\beta_2} & \text{if } \beta_2 \text{ points to } n-2, \\ \ker \varphi_{\beta_2} & \text{otherwise.} \end{cases}$$

Then the condition on the three maps $\varphi_{\alpha_i}, \varphi_{\beta_1}$ and φ_{β_2} is that the three one-dimensional subspaces are pairwise distinct. This corresponds to the "generic" situation as opposed to the special case where two (or more) of these subspaces are equal.

Example 3.2. Let Q be the quiver

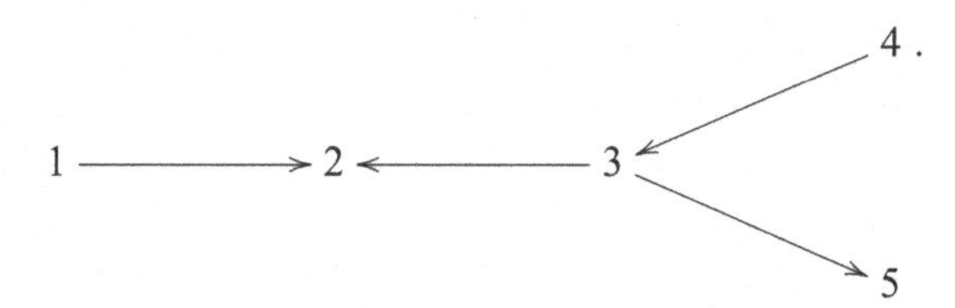

Then

$$P(1) = \begin{smallmatrix} 1 \\ 2 \end{smallmatrix} \quad P(2) = 2 \quad P(3) = \begin{smallmatrix} & 3 & \\ 2 & & 5 \end{smallmatrix}$$

$$P(4) = \begin{smallmatrix} & 4 & \\ & 3 & \\ 2 & & 5 \end{smallmatrix} \quad P(5) = 5$$

and, using the knitting algorithm, the Auslander–Reiten quiver is

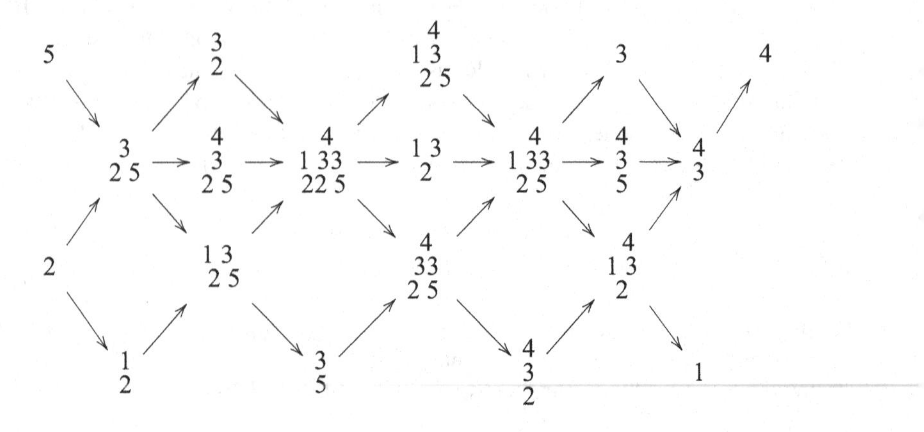

3.3.2 τ-Orbits

As in type A, there are several ways to compute the τ-orbits.

3.3.2.1 First Method: Auslander–Reiten Translation

Let us compute $\tau^{-1}M$ for the module $M = \begin{smallmatrix}1\,3\\2\,5\end{smallmatrix}$ in Example 3.2. The upper line in the following diagram shows an injective resolution of M, and the lower line shows the projective resolution of $\tau^{-1}M$ obtained by applying ν^{-1}.

$$\begin{array}{ccccccccc}
0 & \longrightarrow & \begin{smallmatrix}1\,3\\2\,5\end{smallmatrix} & \longrightarrow & \begin{smallmatrix}4\\1\,3\\2\end{smallmatrix} \oplus \begin{smallmatrix}4\\3\\5\end{smallmatrix} & \longrightarrow & \begin{smallmatrix}4\\3\end{smallmatrix} \oplus 4 & & \\
 & & & & \Big\downarrow \nu^{-1} & & \Big\downarrow \nu^{-1} & & \\
 & & & & 2 \oplus 5 & \longrightarrow & \begin{smallmatrix}3\\2\,5\end{smallmatrix} \oplus \begin{smallmatrix}4\\3\\2\,5\end{smallmatrix} & \longrightarrow & \begin{smallmatrix}4\\33\\2\,5\end{smallmatrix} \longrightarrow 0
\end{array}$$

Thus $\tau^{-1}M = \begin{smallmatrix}4\\33\\2\,5\end{smallmatrix}$ which verifies the result of Example 3.2.

3.3.2.2 Second Method: Coxeter Functor

As in Sect. 3.1.2.2, we define a sequence of vertices $(i_1, i_2, \ldots, i_n)$, with $i_j \neq i_\ell$, if $j \neq \ell$, as follows.

i_1 is a sink of Q.
i_2 is a sink of the quiver $s_{i_1} Q$ obtained from Q by reversing all arrows that are incident to the vertex i.
i_k is a sink of $s_{i_{k-1}} \ldots s_{i_2} s_{i_1} Q$, for $k = 2, 3, \ldots, n$.

Then we define the Coxeter element $c = s_{i_1} s_{i_2} \cdots s_{i_n}$ as a product of reflections using this sequence of vertices.

Thus in Example 3.2, we can take the sequence $(2, 5, 1, 3, 4)$, and its Coxeter element is $c = s_2 s_5 s_1 s_3 s_4$.

Let us use this Coxeter element to compute the dimension vector of $\tau^{-1} \begin{smallmatrix} 1\,3 \\ 2\,5 \end{smallmatrix}$ in Example 3.2. We have $\underline{\dim}\, M = (1, 1, 1, 0, 1)$. Thus $\underline{\dim}\, \tau^{-1} M$ is equal to

$$\begin{aligned} s_2 s_5 s_1 s_3 s_4 (e_1 + e_2 + e_3 + e_5) &= s_2 s_5 s_1 s_3 (e_1 + e_2 + e_3 + e_4 + e_5) \\ &= s_2 s_5 s_1 (e_1 + e_2 + 2e_3 + e_4 + e_5) \\ &= s_2 s_5 (e_2 + 2e_3 + e_4 + e_5) \\ &= s_2 (e_2 + 2e_3 + e_4 + e_5) \\ &= e_2 + 2e_3 + e_4 + e_5 \end{aligned}$$

which again confirms the result obtained in Example 3.2.

As in type $\mathbb{A}$, we can also use the Cartan matrix C and the Coxeter matrix $\Phi = -C^t C^{-1}$ in order to compute the action of the Coxeter element. In our example, we have

$$C = \begin{bmatrix} 1 & 0 & 0 & 0 & 0 \\ 1 & 1 & 1 & 1 & 0 \\ 0 & 0 & 1 & 1 & 0 \\ 0 & 0 & 0 & 1 & 0 \\ 0 & 0 & 1 & 1 & 1 \end{bmatrix} \quad (C^{-1}) = \begin{bmatrix} 1 & 0 & 0 & 0 & 0 \\ -1 & 1 & -1 & 0 & 0 \\ 0 & 0 & 1 & -1 & 0 \\ 0 & 0 & 0 & 1 & 0 \\ 0 & 0 & -1 & 0 & 1 \end{bmatrix}$$

$$\Phi = \begin{bmatrix} 0 & -1 & 1 & 0 & 0 \\ 1 & -1 & 1 & 0 & 0 \\ 1 & -1 & 1 & 1 & -1 \\ 1 & -1 & 1 & 0 & -1 \\ 0 & 0 & 1 & 0 & -1 \end{bmatrix} \quad \Phi^{-1} = \begin{bmatrix} -1 & 1 & 0 & 0 & 0 \\ -1 & 1 & 0 & -1 & 1 \\ 0 & 1 & 0 & -1 & 1 \\ 0 & 0 & 1 & -1 & 0 \\ 0 & 1 & 0 & -1 & 0 \end{bmatrix}.$$

Thus for the representation M above, we can compute the dimension vector of $\tau^{-1} M$ as $\Phi^{-1} \underline{\dim}\, M = \Phi^{-1}(1, 1, 1, 0, 1)^t = (0, 1, 2, 1, 1)^t$.

On the other hand, τM has dimension vector $\Phi(1, 1, 1, 0, 1)^t = (0, 1, 0, 0, 0)^t$.

3.3.3 Arcs of a Punctured Polygon with n Vertices

In this section, we give a geometric construction of the Auslander–Reiten quiver of a quiver Q of type D_n similar to the construction in Sect. 3.1.3. Instead of a triangulated polygon, we work with a triangulated punctured polygon. Thee diagonals in the polygon must be replaced by certain curves that are called *arcs* in the puncture polygon. If the boundary of the polygon has n vertices, then we have exactly n^2 arcs given as follows:

For every vertex a on the boundary of the polygon, we have the $n-2$ arcs shown in the left picture of Fig. 3.8, and for the puncture, we have the n arcs shown in the middle and the n arcs shown in the right picture of Fig. 3.8. Note that for each boundary vertex a, there are two arcs from a to the puncture, and we use a little tag on the arc to distinguish them. The arcs at the puncture that have a tag are called *notched* and the ones without a tag are called *plain*.

Also note that, given two boundary vertices $a \neq b$, there is exactly one arc connecting a and b if a and b are neighbors on the boundary and exactly two arcs if a and b are not neighbors, see Fig. 3.9.

Contrary to the case of the diagonals in the polygon, it is not so straightforward to say when two arcs γ and γ' in the punctured polygon cross.

We denote the number of crossings by $e(\gamma, \gamma')$. If one of the two arcs has both endpoints on the boundary of the polygon, the number of crossing should be

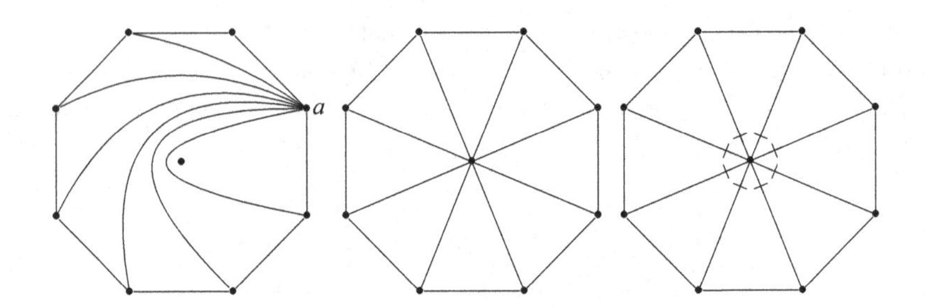

Fig. 3.8 Arcs in a punctured polygon with eight boundary vertices

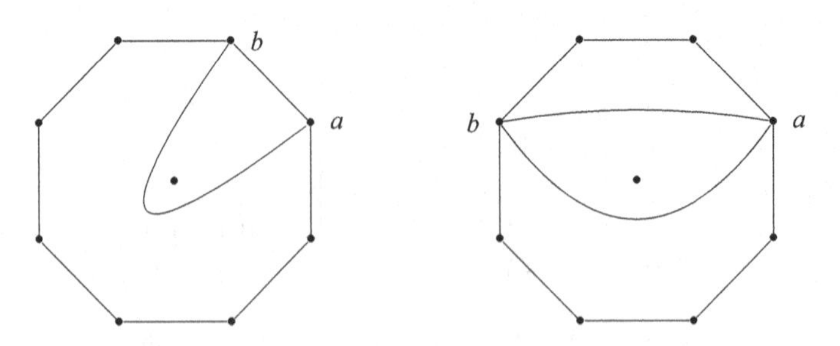

Fig. 3.9 Arcs with specified endpoints

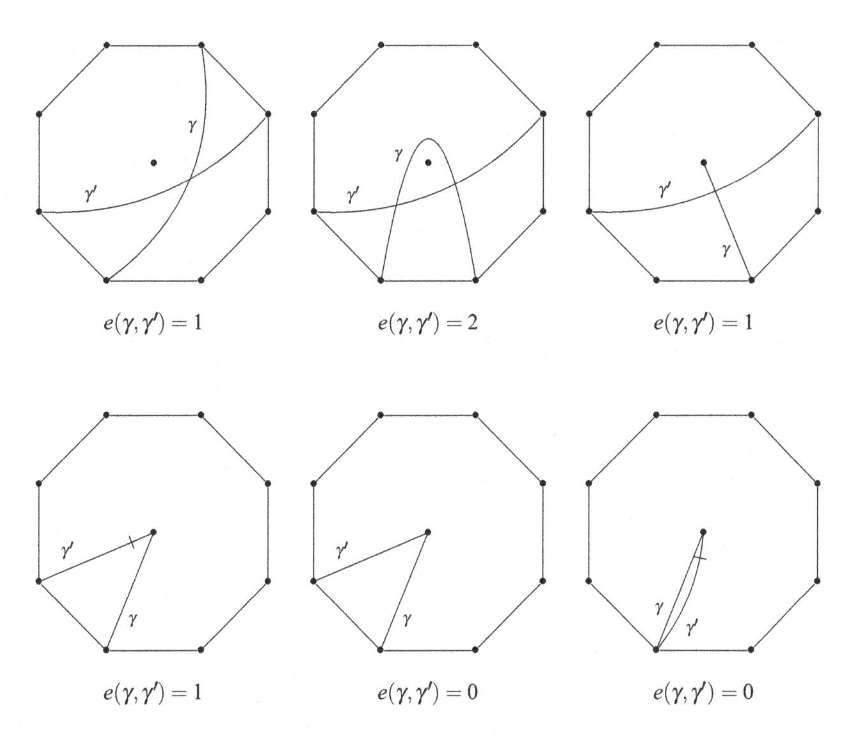

Fig. 3.10 Crossing numbers

intuitively clear, and we show several examples in Fig. 3.10. Note that in this case $e(\gamma, \gamma')$ can be 0, 1, or 2. For a rigorous definition of crossing numbers we would need the notion of homotopy, which would take us too far away from the subject of this book.

If both arcs γ and γ' are incident to the puncture and a and a' denote their respective endpoints on the boundary, we define

$$e(\gamma, \gamma') = \begin{cases} 0 \text{ if } \gamma \text{ and } \gamma' \text{ are both plain,} \\ 0 \text{ if } \gamma \text{ and } \gamma' \text{ are both notched,} \\ 0 \text{ if } a = a', \\ 1 \text{ if } \gamma, \gamma' \text{ have opposite tagging and } a \neq a'. \end{cases}$$

We say that two arcs cross if their crossing number is at least 1, and a *triangulation* is a maximal set of non-crossing diagonals. A triangulation does not necessarily cut the polygon into triangles, even if one allows triangles to have curved edges. Some triangulations are shown in Fig. 3.11.

Now let Q be a quiver of Dynkin type $\mathbb{D}_n$. We associate a triangulation T_Q to Q as follows: Start with an arc γ_1 that cuts off a triangle Δ_0. If $1 \leftarrow 2$ is in Q, then let γ_2 be the unique arc that forms a triangle Δ_1 together with γ_1 and a boundary segment in such a way that γ_1 is counterclockwise from γ_2 in Δ_1. If on the other

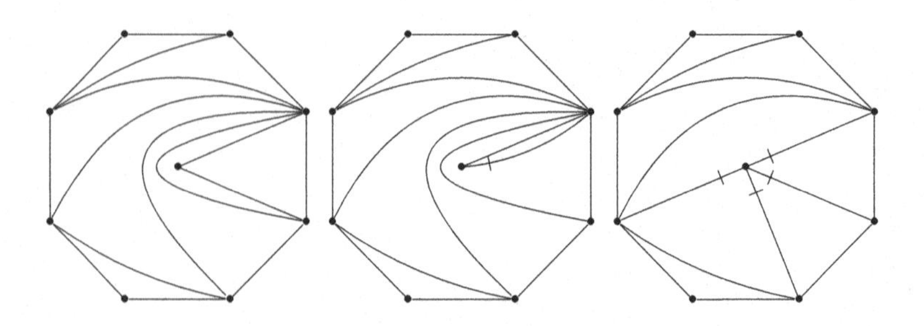

Fig. 3.11 Examples of triangulations

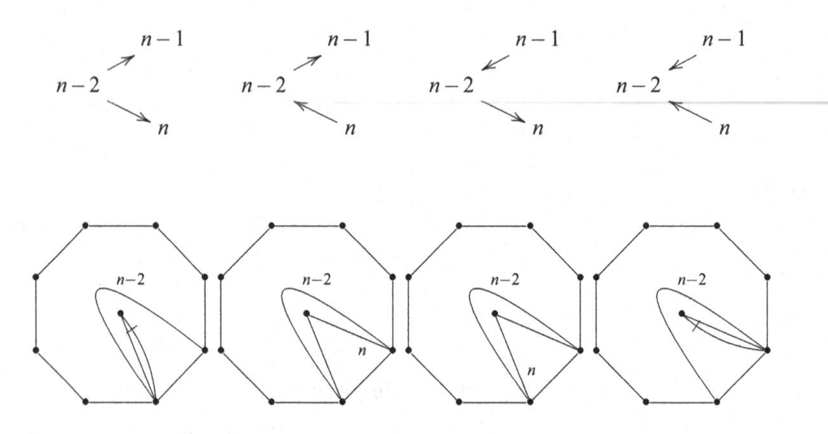

Fig. 3.12 Construction of the triangulation from the quiver

hand, $1 \to 2$ is in Q, then let γ_2 be the unique arc that forms a triangle Δ_1 together with γ_1 and a boundary segment in such a way that γ_1 is clockwise from γ_2 in Δ_1. Continue in this way until $n-2$ arcs are determined. For the arcs γ_{n-1} and γ_n which are corresponding to the vertices $n-1$ and n, respectively, there are four possibilities depending on the orientations of the arrows in the quiver; these four possibilities are displayed in Fig. 3.12.

In this way, the quiver

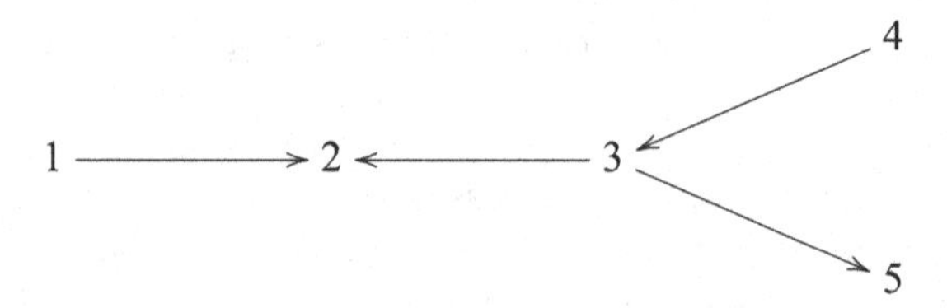

of Example 3.2 gives rise to the triangulation

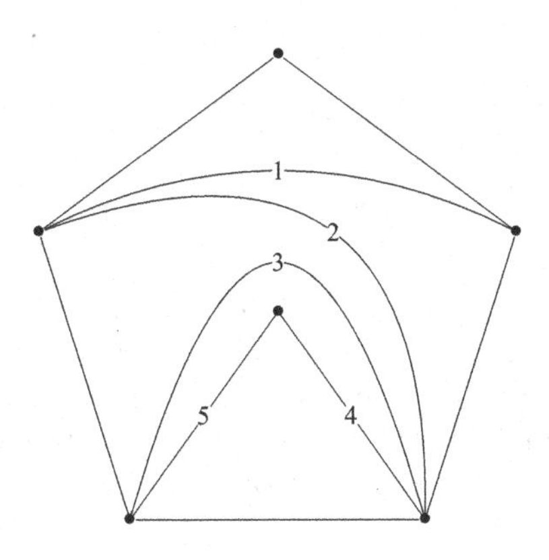

Since T_Q is a triangulation of the punctured polygon, any arc γ which is not already in T_Q will cut through a certain number of diagonals in T_Q; in fact, any such arc γ is uniquely determined by the set of diagonals in T_Q that γ crosses. To such a diagonal γ, we associate the indecomposable representation $M_\gamma = (M_i, \varphi_\alpha)$ of Q whose dimension at vertex i is given by the number of crossings $e(\gamma, \gamma_i)$ between the arc γ and the arc γ_i of the triangulation that corresponds to the vertex i of the quiver. In the example, the arc crosses the arcs $1, 4, 5$ once and $2, 3$ twice, and the corresponding representation is isomorphic to

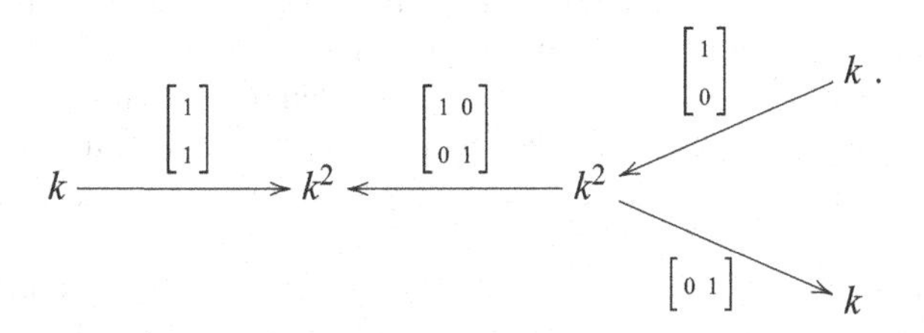

The map $\gamma \mapsto M_\gamma$ is a bijection from the set of arcs that are not in T_Q and the set of isoclasses of indecomposable representations of Q.

The Auslander–Reiten translation τ is given by an elementary clockwise rotation of the punctured polygon with simultaneous change of the tags at the puncture. So in our example

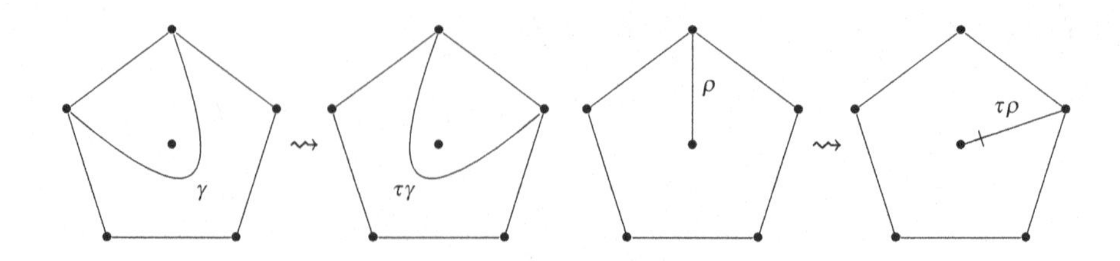

The projective representation $P(i)$ is given by τ^{-1} of the arc γ_i, and the injective representation $I(i)$ is given by τ of the arc γ_i. The complete Auslander–Reiten quiver can be easily constructed now, starting with the projectives and applying the elementary rotation to compute the τ-orbits until we reach the injective in each τ-orbit. The Auslander–Reiten quiver of Example 3.2 is shown in Fig. 3.13.

3.3.4 Computing Hom Dimensions, Ext Dimensions, and Short Exact Sequences

As in type $\mathbb{A}$, we can compute the dimensions of the Hom and Ext spaces easily from the Auslander–Reiten quiver in type $\mathbb{D}$.

3.3.4.1 Dimension of Hom(M, N)

Let Q be a type $\mathbb{D}$ quiver and let M, N be two indecomposable representations of Q. We can compute the dimension of the vector space Hom(M, N) using the relative position of M and N in the Auslander–Reiten quiver. The maximal slanted rectangles of type $\mathscr{A}$ have to be replaced by maximal *hammocks*. It is a little harder to describe these hammocks than the rectangles. Several examples are illustrated in Fig. 3.14.

Recall that a path $M_0 \to M_1 \to \cdots \to M_s$ in the Auslander–Reiten quiver is called a *sectional* path if $\tau M_{i+1} \neq M_{i-1}$ for all $i = 1, \ldots, s-1$. As in type $\mathbb{A}$, we define $\Sigma_{\to}(M)$ to be the set of all indecomposable representations that can be reached from M by a sectional path and $\Sigma_{\leftarrow}(M)$ to be the set of all indecomposable representations from which one can reach M by a sectional path.

We can now construct the hammock by the following algorithm, refer to Fig. 3.14. Start by labeling each vertex in $\Sigma_{\to}(M)$ with the number 1. Then consider the almost split sequence $0 \to M \to E \to \tau^{-1}M \to 0$. Note that each summand of E lies in $\Sigma_{\to}(M)$ and that $\tau^{-1}M$ does not. Label the vertex $\tau^{-1}M$ by the number of indecomposable summands of E minus the label of M. Thus the label at $\tau^{-1}M$ is either $0, 1$ or 2 depending on whether the mesh in the Auslander–Reiten quiver between M and $\tau^{-1}M$ has $1, 2$ or 3 middle vertices, respectively.

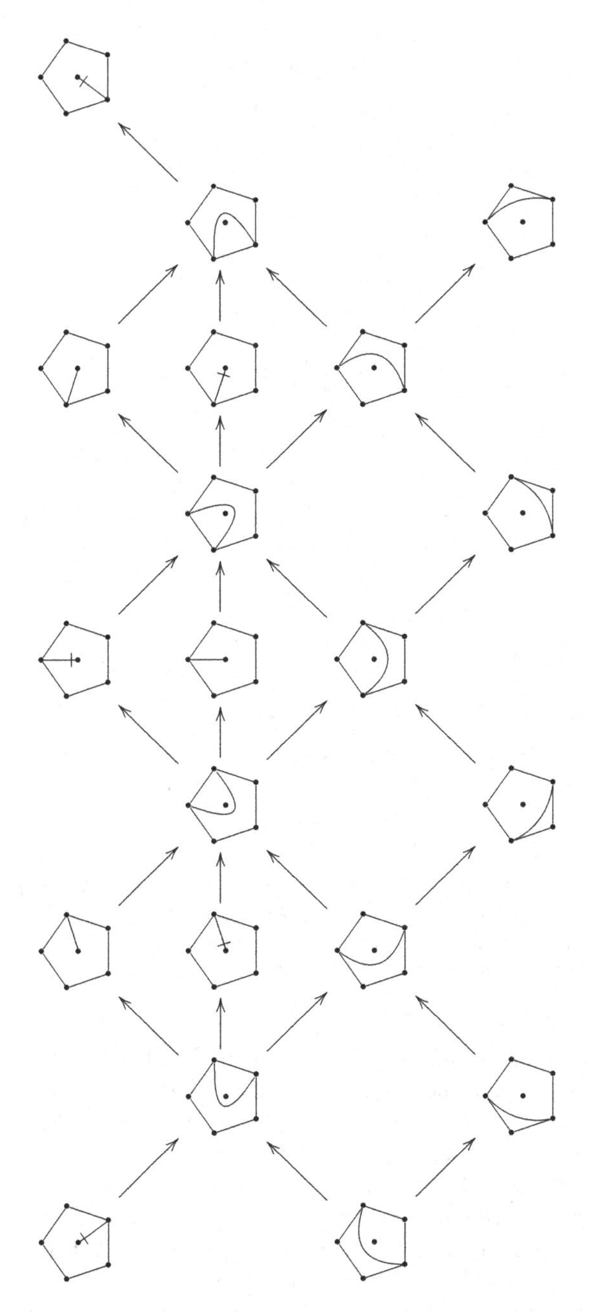

Fig. 3.13 Auslander–Reiten quiver of type $\mathbb{D}_5$ in terms of arcs in a punctured polygon

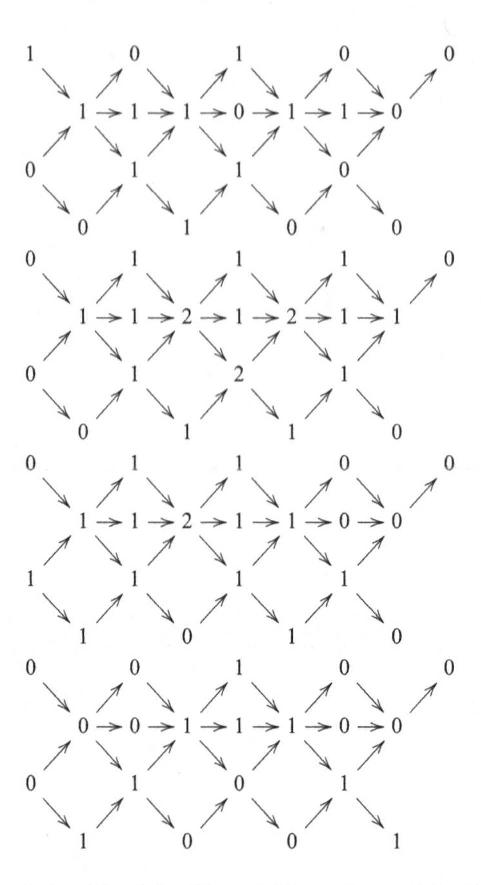

Fig. 3.14 Dimension of Hom$(M,-)$ for $M = P(5)$ on the *top left*, $M = P(3)$ on the *top right*, $M = P(2)$ on the *bottom left*, and $M = P(1)$ on the *bottom right*. The position of the representation M is at the leftmost 1 in each case; the numbers $0, 1, 2$ indicate the dimension of Hom(M, N) for each indecomposable representation N

Recursively, for every almost split sequence $0 \to M' \to E' \to \tau^{-1}N' \to 0$ such that the vertices corresponding to M' and to each summand of E' are already labeled, define the label of the vertex corresponding to $\tau^{-1}M'$ to be the sum of the labels of the indecomposable summands of E' minus the label of M'. If this number is negative, then use the label 0 instead.

This labeling is called the hammock starting at M. If N is any indecomposable representation, then the label at the vertex corresponding to N is the dimension of Hom(M, N). Thus these dimensions can be 0, 1 or 2.

Note that the same algorithm applied to an Auslander–Reiten quiver of type $\mathbb{A}$ will produce the maximal slanted rectangle $\mathscr{R}_{\to}(M)$. Note also that, as in type $\mathbb{A}$, the left boundary of the area with nonzero labels is $\Sigma_{\to}(M)$, and if $M = P(i)$ is an indecomposable projective, then the right boundary of the area with nonzero labels is $\Sigma_{\leftarrow}(I(i))$, and thus the hammock consists of all modules that are nonzero at the vertex i.

3.3.4.2 Ext^1 and Short Exact Sequences

We can compute Ext^1 as in type $\mathbb{A}$ thanks to the formula

$$\dim \operatorname{Ext}^1(M, N) = \dim \operatorname{Hom}(N, \tau M).$$

Thus the dimension of $\operatorname{Ext}^1(M, -)$ is determined by the maximal hammock ending at τM.

Since the dimension of $\operatorname{Ext}^1(M, N)$ can be as large as 2, it is not so easy to find the short exact sequences that represent the elements of $\operatorname{Ext}^1(M, N)$. We know that each element can be represented by short exact sequences of the form $0 \to N \to E \to M \to 0$, where E is some representation of Q, but there might be several choices for E. In the example in Fig. 3.15, there are four non-split short exact sequences starting at N and ending at M, namely

$$\begin{array}{c}
0 \to N \to E_1 \oplus E_2 \oplus H_2 \to M \to 0 \\
0 \to N \to F_1 \oplus F_2 \oplus H_2 \to M \to 0 \\
0 \to N \to \quad G_1 \oplus G_2 \quad \to M \to 0 \\
0 \to N \to \quad H_1 \oplus H_2 \quad \to M \to 0.
\end{array}$$

It is important to note that while there are four non-split short exact sequences, the dimension of $\operatorname{Ext}^1(M, N)$ is only two. Thus any two of the above sequences span the vector space $\operatorname{Ext}^1(M, N)$.

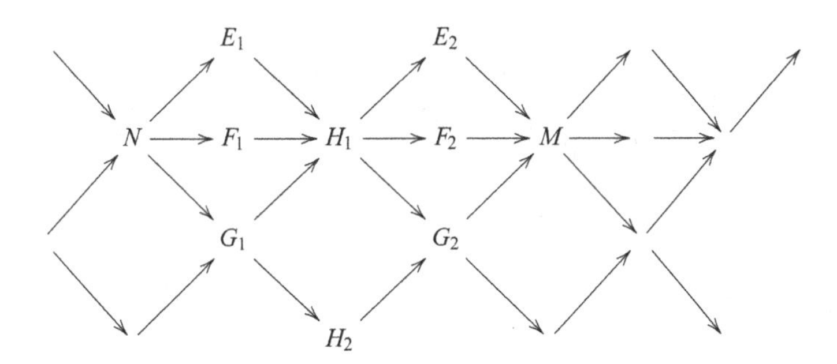

Fig. 3.15 Computing short exact sequences in type $\mathbb{D}$

3.4 Representations of Bound Quivers: Quivers with Relations

In this section, we want to study representations of quivers which, in contrast to earlier sections, are allowed to have oriented cycles or even loops. We had to exclude quivers with oriented cycles in Sects. 2.1–3.3 in order to be able to describe the indecomposable projective representation $P(i)$ at vertex i in terms of the paths that start at i. If the quiver has an oriented cycle that contains the vertex i, then there exist infinitely many paths that start at i, simply because we can run through the oriented cycle over and over again.

For this reason, we will only consider representations that satisfy certain relations given in terms of paths in the quiver. As an example, consider the quiver

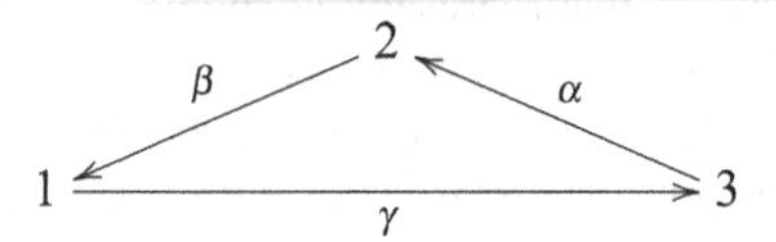

There are infinitely many paths in Q, for example, those starting at vertex 3 include $e_3, \alpha, \alpha\beta, \alpha\beta\gamma, \alpha\beta\gamma\alpha, \alpha\beta\gamma\alpha\beta, \ldots$. We will allow only finitely many paths, by imposing certain *relations*, for example,

$$\alpha\beta = 0, \beta\gamma = 0 \text{ and } \gamma\alpha = 0.$$

Then there are only six nonzero paths, namely $e_1, e_2, e_3, \alpha, \beta$ and γ. Among the representations $M = (M_i, \varphi_\alpha)$ of Q we will then consider only those that satisfy the relations imposed on the quiver, which, in our example, means that $\varphi_\alpha \circ \varphi_\gamma = 0, \varphi_\beta \circ \varphi_\alpha = 0$, and $\varphi_\gamma \circ \varphi_\beta = 0$. For instance, the representation

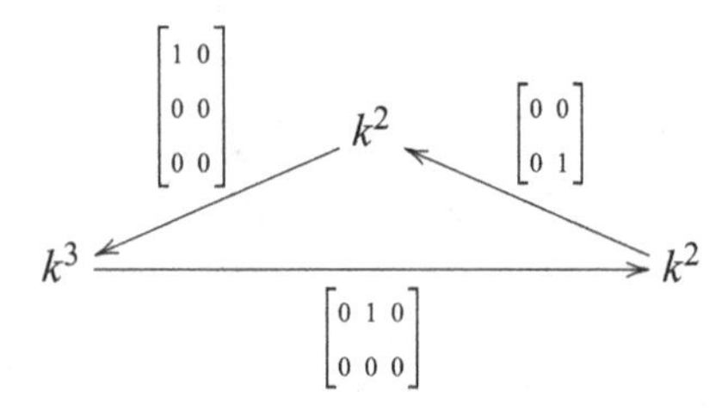

satisfies these relations.

We will now formalize these ideas.

Definition 3.1. Let Q be a quiver.

1. Two paths c, c' in Q are called **parallel** if $s(c) = s(c')$ and $t(c) = t(c')$.
2. A **relation** ρ is a linear combination $\rho = \sum_c \lambda_c c$ of parallel paths each of which has length at least two.

3. A **bound quiver** (Q, R) is a quiver Q together with a set of relations R.

Definition 3.2. Let (Q, R) be a bound quiver. A **representation of** (Q, R) is a representation $M = (M_i, \varphi_\alpha)$ of Q such that $\varphi_\rho = 0$, for each relation $\rho \in R$, where $\varphi_\rho = \sum_c \lambda_c \varphi_c$ if $\rho = \sum_c \lambda_c c$.

Define rep (Q, R) to be the category of representations of (Q, R). We can define morphisms, direct sums, kernels, and cokernels in the same way as in rep Q. The simple representations $S(i)$ are defined in the same way as in rep Q.

To define the indecomposable projective and the indecomposable injective representations, we need the notion of path algebra which we will define in Chap. 4. For now, let us content ourselves with some examples.

Let Q be the quiver

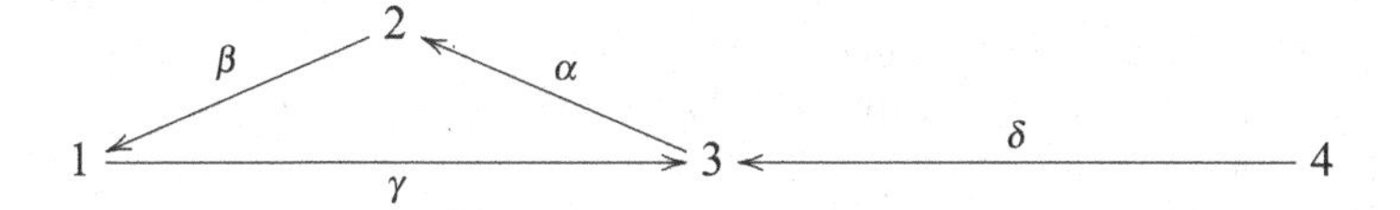

and let $R = \{\alpha\beta, \beta\gamma, \gamma\alpha\}$. Then the paths in the bound quiver (Q, R) are e_1, γ, e_2, β, e_3, α, e_4, δ, $\delta\alpha$, and the indecomposable projective representations are

$$P(1) = \begin{smallmatrix} 1 \\ 3 \end{smallmatrix} \quad P(2) = \begin{smallmatrix} 2 \\ 1 \end{smallmatrix} \quad P(3) = \begin{smallmatrix} 3 \\ 2 \end{smallmatrix} \quad P(4) = \begin{smallmatrix} 4 \\ 3 \\ 2 \end{smallmatrix}.$$

Note that the category rep (Q, R) is not hereditary. Indeed, the simple representation $S(3) = 3$ has the following minimal projective resolution

$$\cdots \longrightarrow \begin{smallmatrix} 3 \\ 2 \end{smallmatrix} \longrightarrow \begin{smallmatrix} 1 \\ 3 \end{smallmatrix} \longrightarrow \begin{smallmatrix} 2 \\ 1 \end{smallmatrix} \longrightarrow \begin{smallmatrix} 3 \\ 2 \end{smallmatrix} \longrightarrow 3 \longrightarrow 0$$

which does not stop after two steps.

3.4.1 Cluster-Tilted Bound Quivers of Type $\mathbb{A}_n$

In Sect. 3.1.3, we have used triangulations of a polygon with $n + 3$ vertices to construct the Auslander–Reiten quiver of the type A quivers. Note however that the triangulations we used then had the property that each triangle has at least one side on the boundary of the polygon. The cluster-tilted quivers of type $\mathbb{A}_n$ are precisely those that are associated to an arbitrary triangulation of the $(n + 3)$-gon.

Let $T = \{1, 2, \ldots, n\}$ be a triangulation of a polygon with $n + 3$ vertices. Define a quiver $Q = (Q_0, Q_1)$ by $Q_0 = T$, and there is an arrow $i \to j$ in Q_1 precisely if the diagonals i and j bound a triangle in which j lies counterclockwise of i:

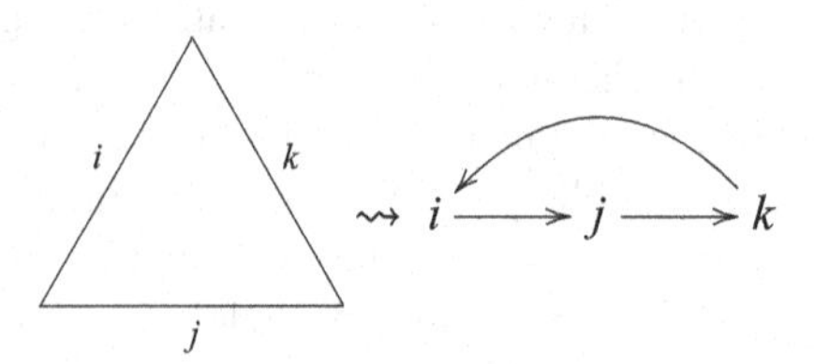

Define the set of relations R to be the set of all paths $i \to j \to k$ such that there exists an arrow $k \to i$. The Auslander–Reiten quiver of (Q, I) can be constructed using diagonals in a polygon with $(n + 3)$ vertices in exactly the same way as for the path algebras of type $\mathbb{A}_n$.

We illustrate this method in an example. Let Q be the quiver

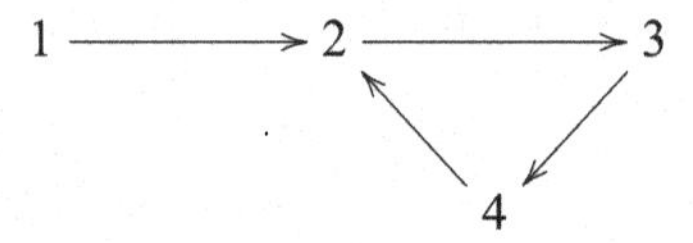

associated to the triangulation

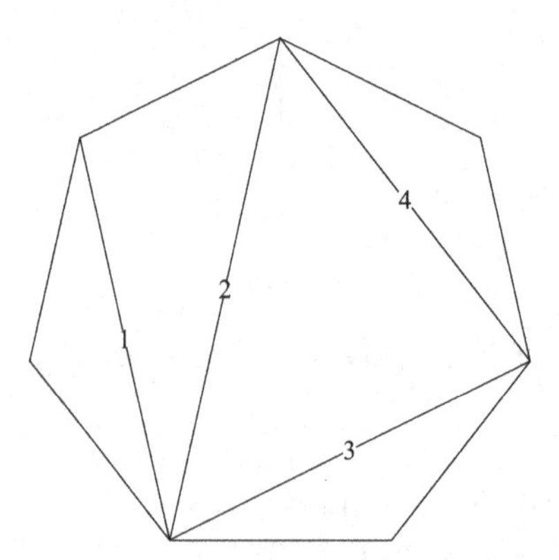

Then the Auslander–Reiten quiver is

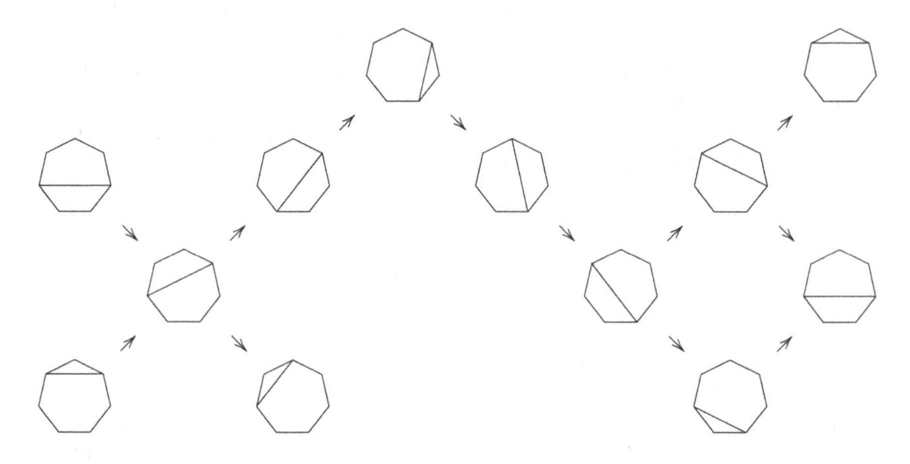

which translates into

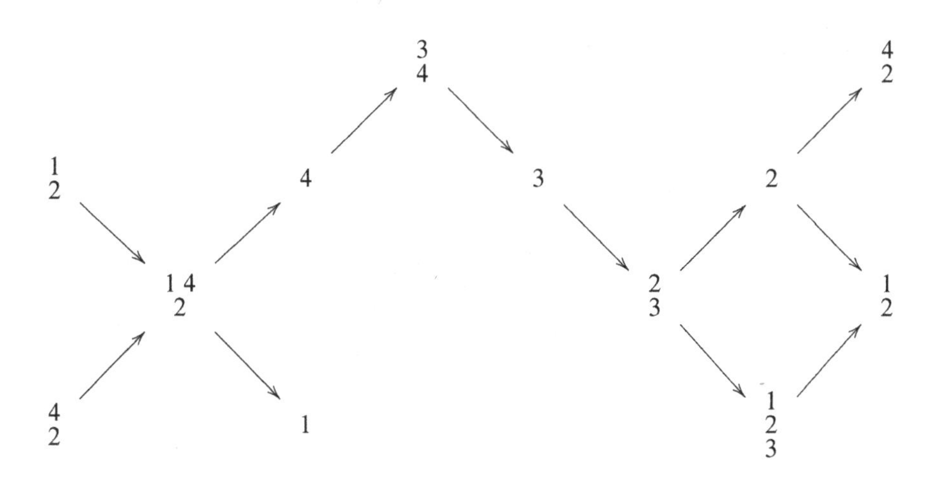

where one has to identify the two representations labeled $\begin{smallmatrix}1\\2\end{smallmatrix}$ and the two representations labeled $\begin{smallmatrix}4\\2\end{smallmatrix}$, so that the Auslander–Reiten quiver has the shape of a Moebius strip.

Note that the number of indecomposable representations of Q is equal to the number of all diagonals in an $(n + 3)$-gon minus the n diagonals in the given triangulation.

Let us compute the number of diagonals. For every vertex a of the polygon, the diagonals starting at a may end at any vertex of the polygon except at a and at its two neighbors. So there are n diagonals starting at each vertex a. There are $n + 3$ possibilities for the vertex a, but when we consider them all, we count each diagonal exactly twice. Therefore the number of diagonals is $n(n + 3)/2$.

Now the number of indecomposable representations of Q is equal to $n(n+3)/2 - n$ which is equal to $n(n + 1)/2$.

In particular the cluster-tilted quivers of type $\mathbb{A}_n$ and the quivers of type $\mathbb{A}_n$ have the same number of indecomposable representations.

3.4.2 Cluster-Tilted Bound Quivers of Type $\mathbb{D}_n$

In Sect. 3.3.3, we have used triangulations of a punctured polygon to compute the Auslander–Reiten quiver of type $\mathbb{D}_n$ quivers. The triangulations we considered then all had the property that there were always exactly two arcs incident to the puncture and that every triangle in the triangulation had at least one edge on the boundary. The **cluster-tilted** quivers of type $\mathbb{D}_n$ are precisely those that are associated to an arbitrary triangulation of the punctured n-gon. The quiver is determined from the triangulation just as in Sect. 3.3.3.

For example, the triangulation

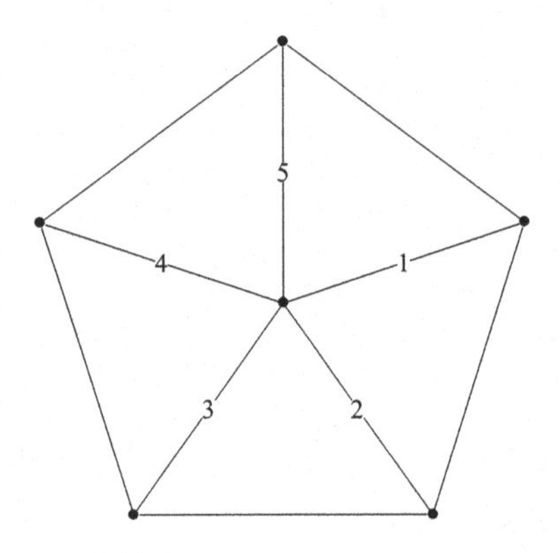

gives rise to the quiver

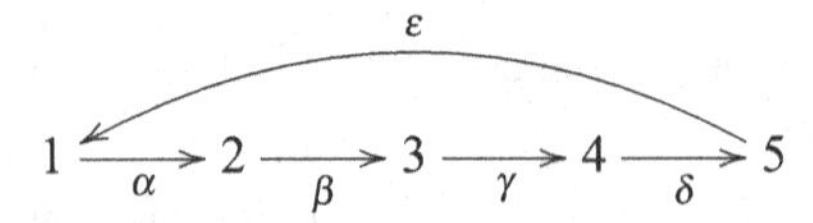

bound by the relations $\alpha\beta\gamma\delta = \beta\gamma\delta\epsilon = \gamma\delta\epsilon\alpha = \delta\epsilon\alpha\beta = \epsilon\alpha\beta\gamma = 0$; and its Auslander–Reiten quiver is given in terms of arcs in Fig. 3.16 and in terms of representations in Fig. 3.17.

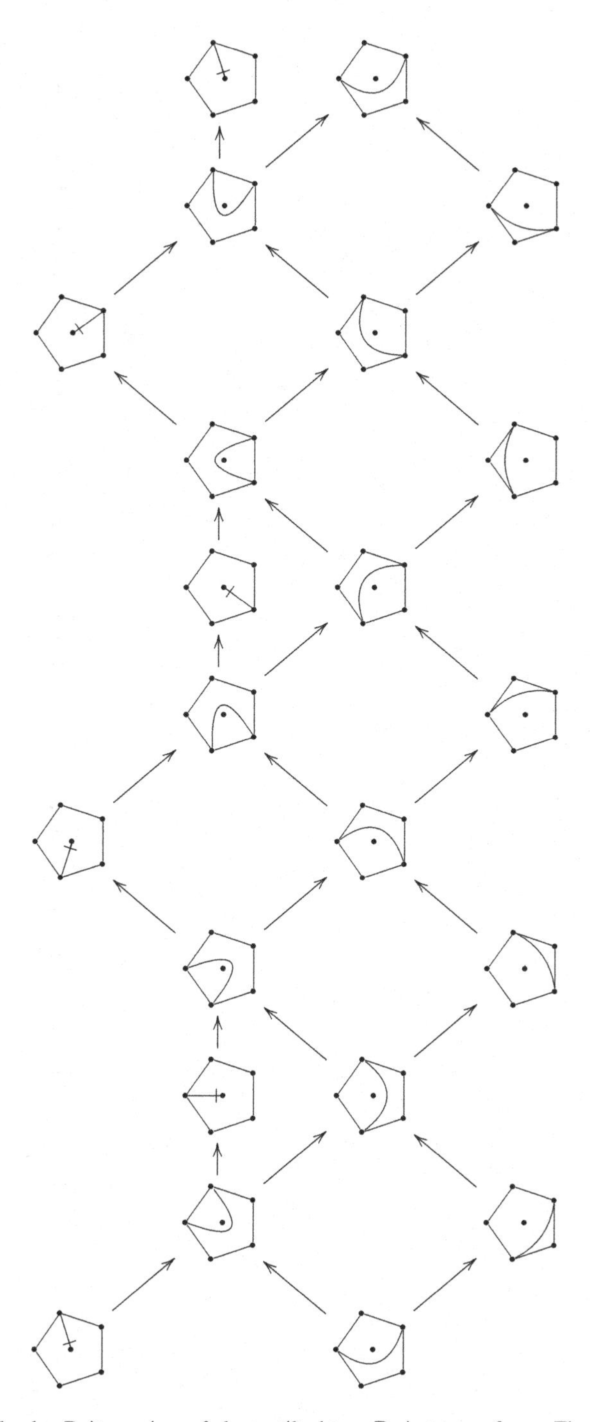

Fig. 3.16 Auslander–Reiten quiver of cluster-tilted type $\mathbb{D}_5$ in terms of arcs. The two vertices on the *far left* are to be identified with the two vertices on the *far right*

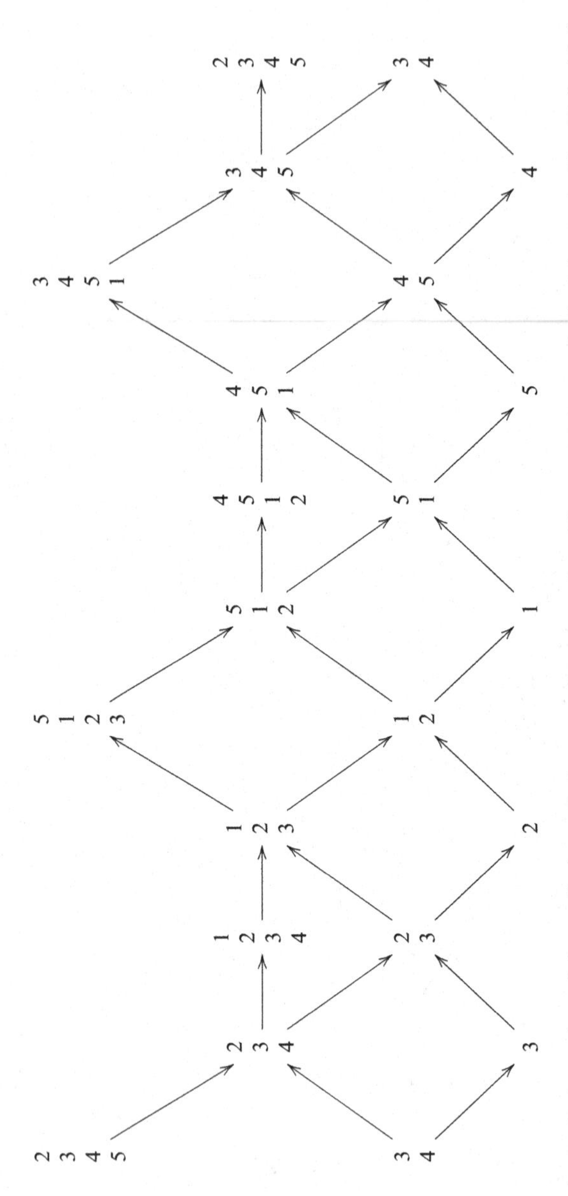

Fig. 3.17 Auslander–Reiten quiver of cluster-tilted type $\mathbb{D}_5$ in terms of representations. The two vertices on the *far left* are to be identified with the two vertices on the *far right*

3.5 Notes

Further information on the construction of Auslander–Reiten quivers can be found in [8, 35]; more on representation type and Gabriel's Theorem in [18, 30, 33]. The construction of Auslander–Reiten quivers from triangulations was introduced in [28, 54].

Problems

Exercises for Chap. 3

3.1. Compute the Auslander–Reiten quivers of the following quivers:

1. $1 \longrightarrow 2 \longleftarrow 3 \longrightarrow 4 \longleftarrow 5 \longleftarrow 6$

2. $1 \longrightarrow 2 \longrightarrow 3 \longrightarrow 4 \longrightarrow 5 \longleftarrow 6$

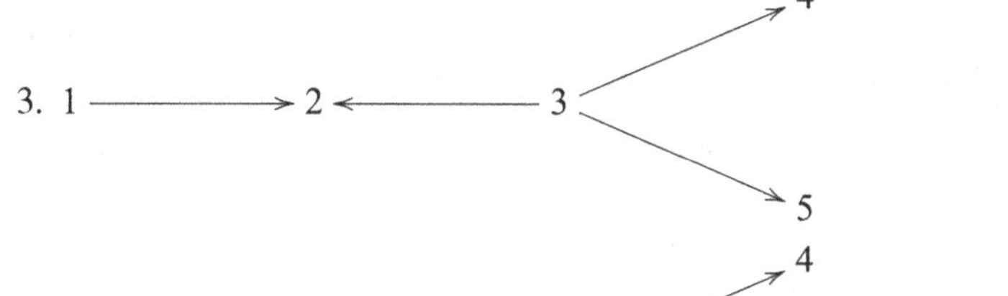

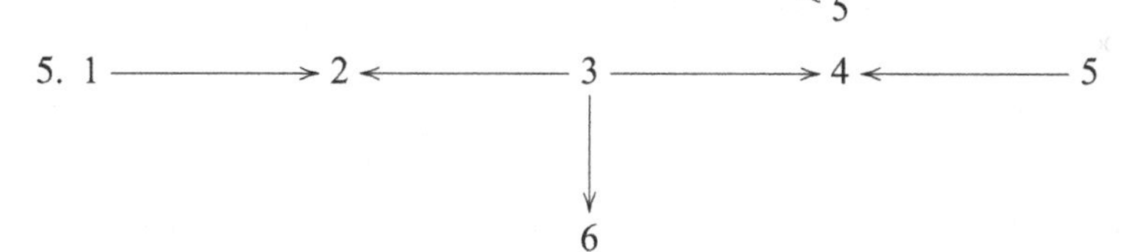

3.2. Let Q be the quiver

$$1 \longrightarrow 2 \longleftarrow 3 \longrightarrow 4 \longleftarrow 5 \longleftarrow 6$$

and consider the indecomposable representations L and N given by the dimension vectors

$$\underline{\dim}\, L = (0, 1, 1, 1, 0, 0) \text{ and } \underline{\dim}\, N = (0, 0, 1, 1, 1, 1).$$

Prove that $\dim \operatorname{Ext}^1(N, L) = 1$ and find the middle term of a non-split short exact sequence of the form

$$0 \longrightarrow L \longrightarrow M \longrightarrow N \longrightarrow 0.$$

3.3. Let Q be the quiver

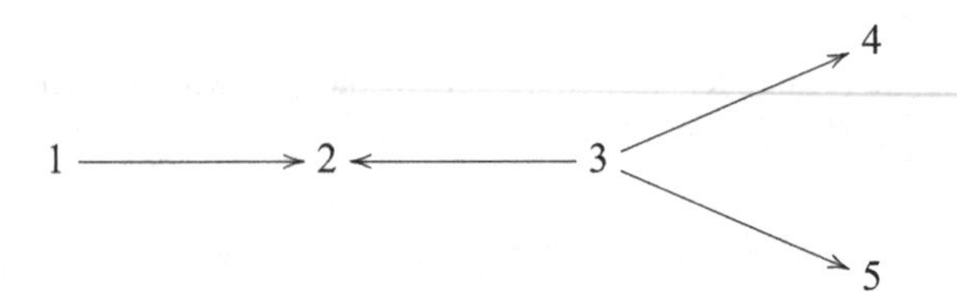

and consider the indecomposable representations L and N given by the dimension vectors $\underline{\dim}\, L = (0, 1, 1, 1, 1)$ and $\underline{\dim}\, N = (1, 1, 1, 0, 0)$.

1. Prove that $\dim \operatorname{Ext}^1(N, L) = 2$ and find 4 non-equivalent non-split short exact sequences of the form

$$0 \longrightarrow L \longrightarrow M \longrightarrow N \longrightarrow 0.$$

2. Show that L is projective and that N is injective.
3. Show that $\tau^3 N$ is a summand of the radical of L.

3.4. Let Q be the quiver

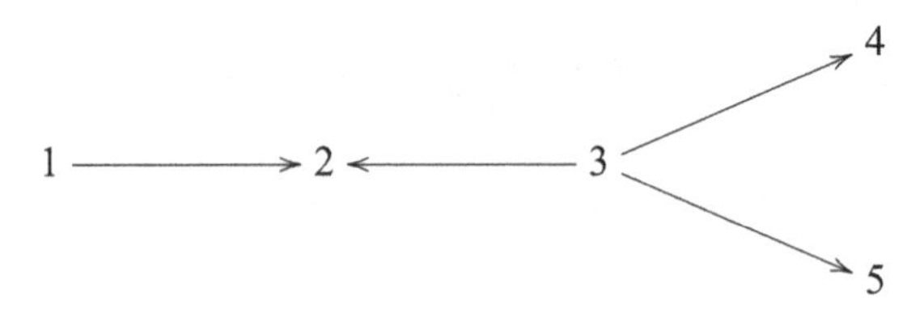

and consider the indecomposable representations $L = \begin{smallmatrix} & 3 \\ 2 & & 4 \end{smallmatrix}$ and $N = \begin{smallmatrix} 1 & & 3 & & 3 \\ & 2 & & 4 & & 5 \end{smallmatrix}$.

1. Prove that there is a unique representation M for which there exists a non-split short exact sequence

$$0 \longrightarrow L \longrightarrow M \stackrel{f}{\longrightarrow} N \longrightarrow 0 .$$

2. Let $M' = \begin{smallmatrix} 1\,3 \\ 2\,5 \end{smallmatrix}$ and $g : M' \to N$ be the inclusion morphism. Then the fiber product X of f and g, defined in Exercise 1.8 of Chap. 1, gives a short exact sequence

$$0 \longrightarrow L \longrightarrow X \longrightarrow M' \longrightarrow 0\,.$$

Prove that $L = \tau M'$.

3.5. Let Q be the quiver

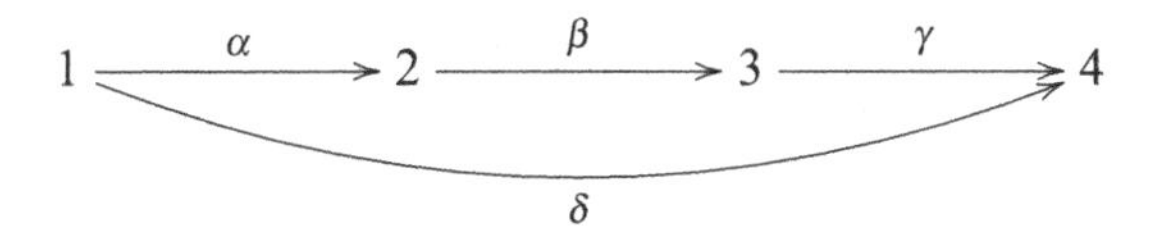

and let M be the indecomposable representation

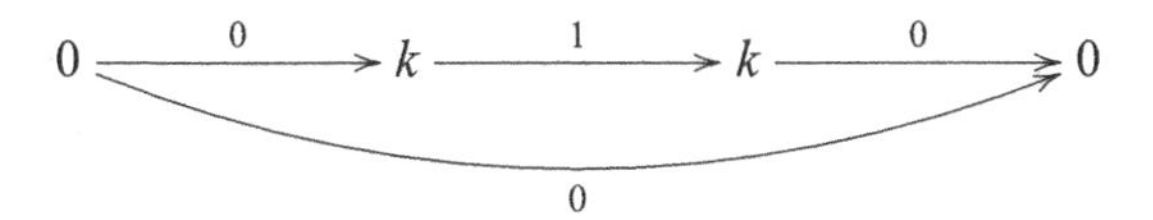

Compute $L_1 = \tau M$, $L_2 = \tau^2 M$ and $L_3 = \tau^3 M$ using the Nakayama functor. Find three representations N_1, N_2 and N_3, by explicitly writing out the matrices, such that $\underline{\dim} N_i = (1, 1, 1, 1)$ and L_i is a subrepresentation of N_i, for $i = 1, 2, 3$.

3.6. Compute the Auslander–Reiten quiver of (Q, I), where Q is the quiver

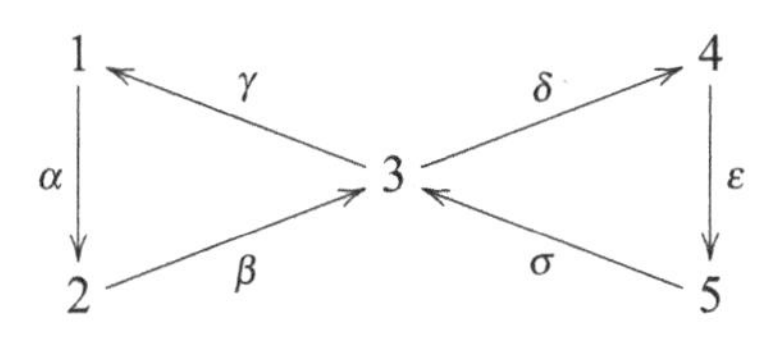

and $I = \{\alpha\beta, \beta\gamma, \gamma\alpha, \delta\epsilon, \epsilon\sigma, \sigma\delta\}$. [Hint: Use a triangulated polygon.]

3.7. Compute the Auslander–Reiten quiver of (Q, I), where Q is the quiver

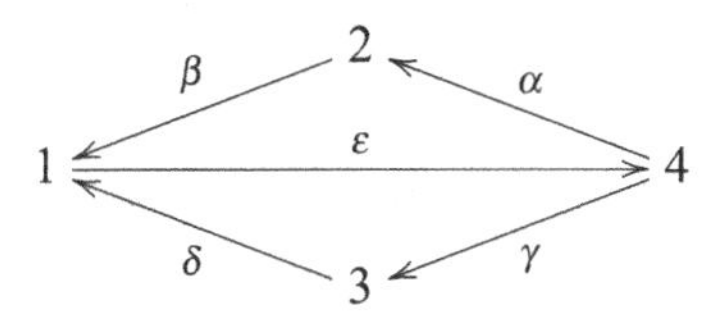

and $I = \{\alpha\beta - \gamma\delta, \epsilon\alpha, \epsilon\gamma, \beta\epsilon, \delta\epsilon\}$. [Hint: Use a triangulated punctured polygon.]

Part II
Path Algebras

In this part of the book, we develop a different approach, regarding quiver representations as modules over a k-algebra A, where k is an algebraically closed field as usual. Such an algebra A has three operations: addition, multiplication, and scaling by elements of k. With respect to addition and scaling, the algebra A is a vector space, and with respect to addition and multiplication, A is a ring.

An A-module M is an abelian group on which the algebra A is acting in such a way that the group operation of M is compatible with the operations of A via this action. A morphism between two modules is a map which preserves the module structure.

Thus every algebra comes with its own module category, and the representation theory of the algebra consists in the study of this category.

On the other hand, every quiver Q defines an algebra, the *path algebra* kQ of Q, simply by taking the set of all paths in the quiver as a basis and defining the multiplication on this basis to be the concatenation of paths. It turns out that the category rep Q of (finite-dimensional) representations of the quiver Q is equivalent to the category mod kQ of (finitely generated) modules over the path algebra kQ.

Moreover, the category of finitely generated modules over *any* finite-dimensional k-algebra is equivalent to the category of finitely generated modules over a quotient of a path algebra.

Chapter 4
Algebras and Modules

This chapter is an introduction to k-algebras and their modules, where k is an algebraically closed field. Since every algebra is a ring, we will often use certain notions from ring theory, like ideals and radicals. We introduce these notions in the first section. In the second and the third section we define k-algebras and their modules and present examples and basic properties. In the fourth section, we study the direct sum decomposition of a k-algebra (as a module over itself) determined by a choice of a complete set of primitive orthogonal idempotents $e_1, \dots, e_n$. For the path algebra of a quiver Q, we are already familiar with this construction, namely the idempotent e_i corresponds to the constant path at the vertex i in Q and the direct sum decomposition of the algebra corresponds to the direct sum of all indecomposable projective representations $P(i)$. In the fifth section, we prove a useful criterion for the indecomposability of a module M. In fact, we show that M is indecomposable if and only if the algebra of all endomorphisms of M is a local algebra.

4.1 Concepts from Ring Theory

Let R be a ring with $1 \neq 0$. A *right ideal* (respectively *left ideal*) I is a subgroup of the additive group of R such that $ar \in I$ (respectively $ra \in I$), for all $a \in I, r \in R$. A *two-sided ideal* is a right ideal that is also a left ideal.

Example 4.1.

1. $0 = \{0\}$ and R are two-sided ideals in R.
2. The kernel of a ring homomorphism is a two-sided ideal. Indeed, let $f : R \to S$ be a ring homomorphism, and let $a \in \ker f$ and $r \in R$. Then $f(ar) = f(a)f(r) = 0f(r) = 0$ and $f(ra) = f(r)f(a) = f(r)0 = 0$, which shows that $ar, ra \in \ker f$. We leave it to the reader to show that $\ker f$ is a subgroup of the additive group R.

R. Schiffler, *Quiver Representations*, CMS Books in Mathematics,
DOI 10.1007/978-3-319-09204-1_4

3. If $a \in R$ then $aR = \{ar \mid r \in R\}$ is a right ideal in R, called the **right ideal generated by** a. On the other hand, $Ra = \{ra \mid r \in R\}$ is a left ideal in R, called the **left ideal generated by** a, and $RaR = \{ras \mid r, s \in R\}$ is a two-sided ideal in R, called the **two-sided ideal generated by** a. The (left, right, or two-sided) ideal generated by a is the smallest (left, right or two-sided) ideal in R that contains a.
4. If $\mathscr{S}$ is a subset of R, then the smallest (left, right, or two-sided) ideal in R that contains $\mathscr{S}$ is called the (left, right, or two-sided) ideal generated by $\mathscr{S}$.
5. If I is a two-sided ideal in the ring R, then the quotient R/I is a ring with respect to the multiplication $(a + I)(a' + I) = aa' + I$.
6. Given a (left, right, or two-sided) ideal I in R and a positive integer m, then

$$I^m = \{\text{finite sums of elements } a_1 a_2 \cdots a_m \mid a_i \in I\}$$

is a (left, right, or two-sided) ideal in R.

Definition 4.1. An ideal I is called **nilpotent** if $I^m = 0$ for some $m \geq 1$.

Definition 4.2. A proper (left, right, or two-sided) ideal I in R is called **maximal** if for any (left, right or two-sided) ideal J such that $I \subset J \subset R$, we have $I = J$ or $J = R$.

In a commutative ring R, an ideal I is maximal if and only if the quotient ring R/I is a field. On the other hand, in a field k, the only ideals are 0 and k.

For us the most important ideal is the following:

Definition 4.3. The (Jacobson) **radical** $\operatorname{rad} R$ is the intersection of all maximal right ideals in R.

It follows from Zorn's Lemma that R contains at least one maximal right ideal, so $\operatorname{rad} R \neq R$.

The radical is an important concept, which we will use frequently. We will show below that the intersection of all maximal *right* ideals is equal to the intersection of all maximal *left* ideals; thus we do not need to distinguish the "left radical" from the "right radical."

Lemma 4.1. *Let R be a ring and let $a \in R$. Then the following are equivalent:*

1. *$a \in \operatorname{rad} R$.*
2. *For all $b \in R$, the element $1 - ab$ has a right inverse.*
3. *For all $b \in R$, the element $1 - ab$ has a two-sided inverse.*
4. *a lies in the intersection of all maximal left ideals in R.*
5. *For all $b \in R$, the element $1 - ba$ has a left inverse.*
6. *For all $b \in R$, the element $1 - ba$ has a two-sided inverse.*

Proof.

($1 \Rightarrow 2$) Let $a \in \operatorname{rad} R$ and $b \in R$. Suppose that $1 - ab$ has no right inverse. Then there exists a maximal right ideal I that contains $1 - ab$, and by definition of the radical, we also have $\operatorname{rad} R \subset I$. Thus $a \in I$, and hence $ab \in I$, because I is a right ideal. But then $1 = 1 - ab + ab \in I$ which implies that $I = R$, a contradiction. Thus $1 - ab$ has a right inverse.

($2 \Rightarrow 3$) Assume that $1 - ab$ has a right inverse that we denote by c. We will show that c is also a left inverse of $1 - ab$. We have $1 = (1 - ab)c$; thus $c = 1 + abc = 1 - a(-bc)$. Now, using (2) again, the element $1 - a(-bc)$ has a right inverse that we denote by d. Then

$$1 = cd = (1 - a(-bc))d = d + abcd = d + ab,$$

which implies that $d = 1 - ab$ and thus $1 = cd = c(1 - ab)$. This shows that c is also a left inverse of $1 - ab$.

($3 \Rightarrow 1$) Suppose that $a \notin \operatorname{rad} R$. Then there exists a maximal right ideal I that does not contain a. Now aR is a right ideal that does contain a, and therefore $I + aR$ is a right ideal that contains a and I, and so $I \subset I + aR$ and $I \neq I + aR$. Since I is a maximal right ideal, it follows that $I + aR = R$. Thus there exist $x \in I$ and $b \in R$ such that $1 = x + ab$, whence $1 - ab = x \in I$. But (3) implies that $1 - ab$ has a right inverse y, thus $1 = xy \in I$, since I is a right ideal, and this implies that $I = R$, a contradiction to the fact that I is a proper ideal.

This shows that (1)–(3) are equivalent. The proof of the equivalence of (4)–(6) is analogous but replacing "right" by "left."

To complete the proof it suffices to show that one of the conditions (1)–(3) is equivalent to one of the conditions (4)–(6). We show that (3)$\Leftrightarrow$(6). If $1 - ab$ has a two sided inverse c, then

$$\begin{aligned} 1 = \quad & 1 - ba + ba && = 1 - ba + b(1 - ab)ca \\ & = 1 - ba + bca - babca = && (1 - ba)(1 + bca). \end{aligned}$$

Thus $(1 + bca)$ is a right inverse of $1 - ba$.

Similarly, $1 = c(1 - ab)$ implies that $1 = (1 + bca)(1 - ba)$, and thus $(1 + bca)$ is also a left inverse of $1 - ba$.

□

Corollary 4.2.

1. $\operatorname{rad} R$ *is the intersection of all maximal left ideals in* R.
2. $\operatorname{rad} R$ *is a two-sided ideal in* R.
3. $\operatorname{rad}(R/\operatorname{rad} R) = 0$.
4. *If* I *is a two-sided nilpotent ideal in* R, *then* $I \subset \operatorname{rad} R$.

Proof. Statements (1) and (2) follow directly from Lemma 4.1.

(3) We need a result from Ring Theory, which says that if J is a two-sided ideal in a ring A, then the map $I \mapsto I/J$ is a bijection between the ideals I in R that contain J and the ideals in R/J. This bijection sends maximal ideals to maximal ideals. Applied to our situation, we see that the maximal

ideals in $R/\mathrm{rad}\, R$ are of the form $I/\mathrm{rad}\, R$ where I is a maximal ideal in R. Consequently, the radical of $R/\mathrm{rad}\, R$ which is the intersection of all maximal ideals in $R/\mathrm{rad}\, R$ is equal to the quotient of the intersection of all maximal ideals in R by the radical of R, that is, $\mathrm{rad}\,(R/\mathrm{rad}\, R) = \mathrm{rad}\, R/\mathrm{rad}\, R = 0$.

(4) Suppose that I is a two-sided nilpotent ideal, and let $m \geq 1$ be such that $I^m = 0$. Let $x \in I$. Then, since $ax \in I$ for all $a \in R$, we have $(ax)^m = 0$ for all $a \in R$. Thus

$$1 = 1 - (ax)^m = (1 + ax + (ax)^2 + \cdots + (ax)^{m-1})(1 - ax),$$

which shows that $1 - ax$ has a left inverse, for all $a \in R$. Lemma 4.1 implies that $x \in \mathrm{rad}\, R$, and therefore, $I \subset \mathrm{rad}\, R$. □

A ring that contains only one maximal right ideal is called a **local ring**. In a local ring the unique maximal right ideal is thus equal to the radical. We will study local algebras in Sect. 4.5

4.2 Algebras

Let k be an algebraically closed field.

Definition 4.4. A k-*algebra* A is a ring $(A, +, \cdot)$ with unity 1 such that A has also a k-vector space structure such that

1. the addition in the vector space A is the same as in the ring A,
2. the scalar multiplication in the vector space A is compatible with the ring multiplication, that is, for all $\lambda \in k$ and all $a, b \in A$, we have

$$\lambda(ab) = (\lambda a)b = a(\lambda b) = (ab)\lambda.$$

The *dimension* of the algebra A is the dimension of the vector space A.

Example 4.2.

1. The ring of polynomials $k[X]$ in one indeterminate X with coefficients in k is an algebra whose unity is the constant polynomial 1. The scalar multiplication by λ is simply given by multiplying each coefficient by λ.
2. The set $\mathrm{Mat}(n, k)$ of $n \times n$ matrices with coefficients in k is an algebra whose unity is the identity matrix. The ring structure is given by the addition and multiplication of matrices, and the scalar multiplication is given by multiplying each entry of a matrix by the given scalar.
3. The set of lower triangular (respectively upper triangular) matrices is a subalgebra of $\mathrm{Mat}(n, k)$. The reason for this is that the identity matrix is triangular, and the set of triangular matrices is closed under matrix addition, matrix multiplication, and scalar multiplication. The invertible elements are the matrices

whose diagonal entries are all nonzero, and the nilpotent elements are the matrices whose diagonal entries are all equal to zero.

4. The set of all 3×3 matrices of the form

$$\begin{bmatrix} * & 0 & 0 \\ 0 & * & 0 \\ * & * & * \end{bmatrix},$$

where each $*$ represents an arbitrary element of k, is an algebra. Again, the identity matrix is of this form and the set is closed under addition, multiplication, and scalar multiplication.

5. The set B of all matrices of the form

$$\begin{bmatrix} * & 0 & 0 \\ * & * & 0 \\ 0 & * & * \end{bmatrix},$$

where each $*$ represents an arbitrary element of k, is not an algebra, because it is not closed under multiplication. For example the product

$$\begin{bmatrix} 0 & 0 & 0 \\ 0 & 0 & 0 \\ 0 & 1 & 0 \end{bmatrix} \begin{bmatrix} 0 & 0 & 0 \\ 1 & 0 & 0 \\ 0 & 0 & 0 \end{bmatrix} = \begin{bmatrix} 0 & 0 & 0 \\ 0 & 0 & 0 \\ 1 & 0 & 0 \end{bmatrix}$$

is not in B.

6. If A is an algebra, then the *opposite algebra* A^{op} is defined on the same underlying vector space by the multiplication rule ab in A^{op} is the same as ba in A.

If $B = \{b_1, b_2, \ldots, b_n\}$ is a basis for the vector space A, then every $a \in A$ is a linear combination of this basis. Now given two such elements $a, a' \in A$, where $a = \lambda_1 b_1 + \lambda_2 b_2 + \cdots + \lambda_n b_n$ and $a' = \lambda'_1 b_1 + \lambda'_2 b_2 + \cdots + \lambda'_n b_n$, their product must satisfy

$$aa' = \sum_{i=1}^{n} \lambda_i b_i \sum_{j=1}^{n} \lambda'_j b_j = \sum_{i,j=1}^{n} \lambda_i \lambda'_j b_i b_j$$

which means that, if we specify how to multiply any two basis elements b_i, b_j, then the multiplication in the algebra is completely determined. We say that we extend the multiplication from the basis to the whole algebra by linearity.

The examples we are mainly interested in are the path algebras of quivers. The definition of path algebras uses the above strategy. Recall that all our quivers are finite.

Given two paths $c = (i|\alpha_1, \alpha_2, \ldots, \alpha_r|j)$ and $c' = (j|\alpha'_1, \alpha'_2, \ldots, \alpha'_{r'}|k)$ with $j = t(c) = s(c')$, we denote by $c \cdot c'$ the concatenation of the two paths; thus

$$c \cdot c' = (i|\alpha_1, \alpha_2, \ldots, \alpha_r, \alpha'_1, \alpha'_2, \ldots, \alpha'_{r'}|k).$$

Definition 4.5. Let Q be a quiver. The **path algebra** kQ of Q is the algebra with basis the set of all paths in the quiver Q and with multiplication defined on two basis elements c, c' by

$$cc' = \begin{cases} c \cdot c' & \text{if } s(c') = t(s) \\ 0 & \text{otherwise.} \end{cases}$$

Thus the product of two arbitrary elements $\sum_c \lambda_c c, \sum_{c'} \lambda'_{c'} c'$ of kQ is given by $\sum_{c,c'} \lambda_c \lambda'_{c'} cc'$.

Lemma 4.3. *In a path algebra kQ, the unity element is given by the sum of all constant paths:*

$$1 = \sum_{i \in Q_0} e_i.$$

Proof. Let $a \in A$. Then $a = \sum_c \lambda_c c$ for some $\lambda_c \in k$. Then $a \sum_{i \in Q_0} e_i = \sum_{i \in Q_0} \sum_c \lambda_c c e_i$, and ce_i is zero if the path c does not end in the vertex i, and $ce_i = c$ if the path c does end in i. Hence $a \sum_{i \in Q_0} e_i = \sum_{i \in Q_0} \sum_{c:t(c)=i} \lambda_c c = \sum_c \lambda_c c = a$. Similarly, one can show that $\sum_{i \in Q_0} e_i a = a$, and the lemma is proved. □

Example 4.3.

1. Let Q be the quiver $1 \circlearrowright \alpha$. The paths in Q are $e_1, \alpha, \alpha^2, \alpha^3, \alpha^4, \ldots$; thus the algebra kQ has basis $\{\alpha^t \mid t = 0, 1, 2, \ldots\}$, and the multiplication is given by $\alpha^s \alpha^t = \alpha^{s+t}$. It follows that kQ is isomorphic to the algebra of polynomials $k[x]$.
2. Let Q be the quiver

$$1 \xrightarrow{\alpha_1} 2 \xrightarrow{\alpha_2} 3 \xrightarrow{\alpha_3} \cdots \xrightarrow{\alpha_{n-1}} n\,.$$

Then kQ is isomorphic to the algebra of upper triangular $n \times n$ matrices, which is readily seen by listing the basis elements as

$$\begin{array}{cccccc}
e_1 & \alpha_1 & \alpha_1\alpha_2 & \alpha_1\alpha_2\alpha_3 & \ldots & \alpha_1\alpha_2\alpha_3\cdots\alpha_{n-1} \\
 & e_2 & \alpha_2 & \alpha_2\alpha_3 & \ldots & \alpha_2\alpha_3\cdots\alpha_{n-1} \\
 & & e_3 & \alpha_3 & \ldots & \alpha_3\cdots\alpha_{n-1} \\
 & & & e_4 & \ldots & \vdots \\
 & & & & \ddots & \vdots \\
 & & & & & e_n
\end{array}$$

3. Let Q be the quiver

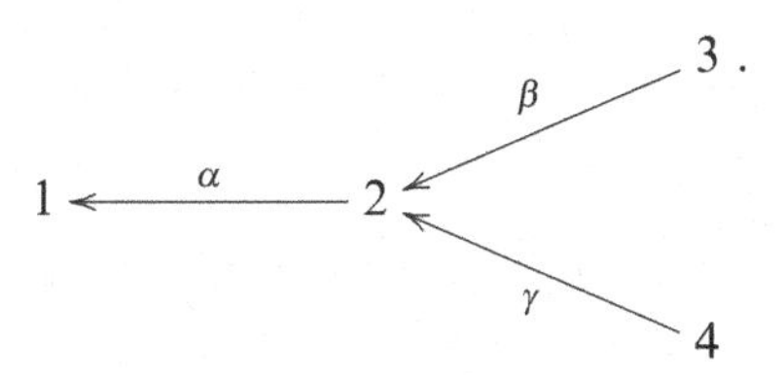

Then kQ has basis $\{e_1, e_2, e_3, e_4, \alpha, \beta, \gamma, \beta\alpha, \gamma\alpha\}$, and kQ is isomorphic to the algebra of matrices of the form

$$\begin{bmatrix} \lambda_{e_1} & 0 & 0 & 0 \\ \lambda_\alpha & \lambda_{e_2} & 0 & 0 \\ \lambda_{\beta\alpha} & \lambda_\beta & \lambda_{e_3} & 0 \\ \lambda_{\gamma\alpha} & \lambda_\gamma & 0 & \lambda_{e_4} \end{bmatrix},$$

where each $\lambda_c \in k$ is the coefficient of the path c. The coefficient at the ith row and jth column in the matrix corresponds to the path from i to j.

Definition 4.6. If A and B are k-algebras, then a k-linear map $f : A \to B$ is a **homomorphism of k-algebras** if $f(1) = 1$ and, for all $a, a' \in A$,

$$f(aa') = f(a)f(a').$$

Definition 4.7. Let B be a k-vector subspace of A. Then B is a **subalgebra** if B contains the unity element 1 and for every $b, b' \in B$ we have $bb' \in B$.

Example 4.4.

1. A is a (two-sided) ideal in A, and A is a subalgebra of A. Moreover, the only ideal I that is also a subalgebra is $I = A$, because being a subalgebra, I contains 1 and, being an ideal, I contains every $1a = a \in A$.
2. If $A = kQ$ and e_i is a constant path, then the right ideal generated by e_i consists of elements $e_i \sum_c \lambda_c c = \sum_c \lambda_c e_i c = \sum_{c:s(c)=i} \lambda_c c$. Thus the set of all paths starting at i is a basis for $e_i A$. Similarly, the set of all paths ending at i is a basis for Ae_i. If α is an arrow in Q, then the right ideal generated by α has as basis the set of paths that start with the arrow α.

3. If $A = kQ$ and e_i, e_j are constant paths, then the right ideal generated by $\{e_i, e_j\}$ has as a basis the set of all paths that start at i or j.
4. Let $A = kQ$ be a path algebra. For $i \in Q_0$, let $\mathscr{S}_i$ be the set of all paths in Q except the constant path e_i. Then the right (or left) ideal generated by $\mathscr{S}_i$ is maximal.

Proposition 4.4. *If I is a two-sided nilpotent ideal in A such that the algebra A/I is isomorphic to a direct product $k \times k \times \cdots \times k$ of copies of the field k, then $I = \operatorname{rad} A$.*

Proof. Corollary 4.2 (4) implies that $I \subset \operatorname{rad} A$; hence it only remains to show that $\operatorname{rad} A \subset I$.

Since k is a field, the only ideals in k are 0 and k, and therefore the maximal ideals in $k \times k \times \cdots \times k$ are

$$0 \times k \times \cdots \times k,\ k \times 0 \times k \times \cdots \times k,\ \ldots,\ k \times k \times \cdots \times k \times 0,$$

which shows that

$$\operatorname{rad}(A/I) = 0. \tag{4.1}$$

Consider the canonical projection $\pi : A \to A/I$ defined by $\pi(a) = a + I$, and let $a \in \operatorname{rad} A$. By Lemma 4.1, we know that for every $b \in A$, the element $1 - ba$ has a two-sided inverse in A that we denote by c. Then $1 + I = \pi(1) = \pi(c(1 - ba)) = \pi(c)\pi(1 - ba) = \pi(c)(1 - \pi(b)\pi(a))$, and thus $1 - \pi(b)\pi(a)$ has a left inverse in A/I. Again by Lemma 4.1, we have $\pi(a) \in \operatorname{rad}(A/I)$, and by (4.1), we get that $\pi(a) = 0$. In other words $a \in I$, and this shows that $\operatorname{rad} A \subset I$. □

Corollary 4.5. *If Q is a quiver without oriented cycles, then* $\operatorname{rad} kQ$ *is the (two sided) ideal generated by all arrows in Q.*

Proof. We denote by R_Q the ideal generated by all arrows in Q. Let ℓ be the largest integer such that Q contains a path of length ℓ. Then any product of $\ell + 1$ arrows is zero; thus $R_Q^{\ell+1} = 0$, which shows that R_Q is a nilpotent ideal. Moreover, the set $\{e_i + R_Q \mid i \in Q_0\}$ is a basis for kQ/R_Q; whence kQ/R_Q is isomorphic to a direct product $\prod_{i \in Q_0} k$. Now Proposition 4.4 implies that $R_Q = \operatorname{rad} kQ$. □

Remark 4.6. For this corollary, we really need the condition that Q has no oriented cycles. For example, for the quiver

$$1 \circlearrowright \alpha$$

the path algebra kQ is isomorphic to the polynomial algebra $k[x]$. Since every linear polynomial $x - a$, with $a \in k$, generates a maximal ideal, we see that $\operatorname{rad} k[x] = 0$.

4.3 Modules

In this section, we introduce modules over a ring R with 1. In later sections, we will be mostly interested in the special case where R is a k-algebra, but since most of the results in this section hold for rings, we present them here in that context.

Definition 4.8. Let R be a ring with $1 \neq 0$. A (right) **R-module** M is an abelian group together with a binary operation, called right R-action,

$$\begin{aligned} M \times R &\longrightarrow M \\ (m, r) &\mapsto mr \end{aligned}$$

such that for all $m_1, m_2 \in M$ and all $r_1, r_2 \in R$, we have

$$\begin{aligned} (1)\ (m_1 + m_2)\, r &= m_1 r + m_2 r, \\ (2)\ m_1(r_1 + r_2) &= m_1 r_1 + m_1 r_2, \\ (3)\quad m_1(r_1 r_2) &= (m r_1) r_2, \\ (4)\quad m_1 1 &= m_1. \end{aligned}$$

One can define *left* R-modules simply by multiplying the elements of the ring from the left and writing the analogous axioms. In this book, the terminology R-module will always mean right R-module.

Example 4.5.

1. The ring R is an R-module with respect to the right R-action given by the multiplication in R.
2. If I is a right ideal in R, then I is an R-module also with respect to the right R-action given by the multiplication in R. In particular, the right ideal generated by a, which is equal to $aR = \{ar \mid r \in R\}$, is an R-module.
3. If I is a right ideal in R and M is a an R-module, then the set

$$MI = \{m_1 r_1 + \cdots + m_t r_t \mid m_i \in M, r_i \in I\}$$

   is a submodule of M.
4. If the ring A is also a k-algebra, then any A-module M is also a k-vector space with respect to $m\lambda = m(\lambda 1_A)$, for $m \in M$ and $\lambda \in k$. This vector space structure is called the *underlying vector space* of the A-module M.

   In practice, this fact is very convenient. For example, it allows us to express the elements of the module as linear combinations of some chosen basis or to prove results by induction on the dimension of a module.
5. If Q is a quiver and $A = kQ$ is its path algebra, then, for each $i \in Q_0$, we can define a module $S(i)$ whose abelian group is the one-dimensional vector space with basis $\{e_i\}$ and whose A-module structure is given on the paths c by

$$me_i c = \begin{cases} me_i & \text{if } c = e_i, \\ 0 & \text{otherwise.} \end{cases}$$

The reader should check that the module axioms are satisfied.

6. If Q is a quiver and $A = kQ$ is its path algebra, then, for each $i \xrightarrow{\alpha} j \in Q_1$, we can define a module $M(\alpha)$ whose vector space is equal to the two-dimensional vector space k^2 with basis $\{e_i, \alpha\}$ and whose A-module structure is given on the paths c by

$$(\lambda_i e_i + \lambda_\alpha \alpha)\, c = \lambda_i e_i c + \lambda_\alpha \alpha c = \begin{cases} \lambda_i e_i & \text{if } c = e_i, \\ \lambda_\alpha \alpha & \text{if } c = e_j, \\ \lambda_i \alpha & \text{if } c = \alpha, \\ 0 & \text{otherwise.} \end{cases}$$

Again, the reader should check that the module axioms are satisfied.

Remark 4.7. The reader will have noticed the similarity between the modules $S(i), M(\alpha)$ and the quiver representations $S(i)$ and $\begin{smallmatrix} i \\ j \end{smallmatrix}$. This is no coincidence as we shall see in Theorem 5.4.

Definition 4.9. A module M is said to be *generated by the elements* $m_1, m_2, \ldots, m_s$ if, for every $m \in M$, there exist $a_i \in R$ such that $m = m_1 a_1 + m_2 a_2 + \cdots + m_s a_s$. The module M is called **finitely generated** if it is generated by a finite set of elements.

Note that if M is generated by $m_1, m_2, \ldots, m_s$, then $M = m_1 R + m_2 R + \cdots + m_s R$. For example, the ideal aR is an R-module generated by one element a.

Definition 4.10. Let M and N be two R-modules. A map $h : M \longrightarrow N$ is called a **morphism** of R-modules if, for all $m, m' \in M$ and all $a \in R$, we have

$$\begin{aligned} h(m + m') &= h(m) + h(m'), \\ h(m\, a) &= h(m)\, a. \end{aligned}$$

The **kernel** of h is the set $\ker h = \{m \in M \mid h(m) = 0\}$, the **image** of h is the set $\operatorname{im} h = \{h(m) \mid m \in M\}$, and the **cokernel** of h is the set $\operatorname{coker} h = N/\operatorname{im} h$.

Note that if A is a k-algebra, then a morphism of A-modules is also a homomorphism of the underlying k-vector spaces, thus a linear map.

Proposition 4.8. *The kernel, image, and cokernel of a morphism are A-modules.*

Proof. Exercise. □

Example 4.6. Let $A = kQ$ be a path algebra, and let $S(j)$ and $M(\alpha)$ be the modules defined in Example 4.5 (4) and (5), where j is a vertex of Q and α is an arrow of Q ending at j. Then there is a morphism $h : S(j) \to M(\alpha)$ defined by $h(me_j) = m\alpha$.

Let us check that h really is a morphism. We have $h(me_j+m'e_j)=h((m+m')e_j)=(m+m')\alpha=m\alpha+m'\alpha=h(me_j)+h(m'e_j)$, and $h(\lambda me_j)=\lambda m\alpha=\lambda h(me_j)$; thus h is k-linear. Moreover $h(me_j\,e_j)=h(me_j)=m\alpha=m\alpha\,e_j=h(me_j)\,e_j$ and if c is a path different from e_j, then $h(me_j\,c)=h(0)=0=m\alpha\,c=h(me_j)\,c$. Thus h is a morphism of A-modules.

Example 4.7. Let A be a k-algebra and M an A-module. An **endomorphism** of M is a morphism of modules from M to M. The set of all endomorphisms of M is denoted by $\operatorname{End} M$. It has a k-vector space structure given by the addition and scalar multiplication of morphisms; and this vector space is an algebra whose multiplication is given by the composition of endomorphisms.

Lemma 4.9 (Nakayama's Lemma). *Let M be a finitely generated R-module and let I be a two-sided ideal in R that is contained in* rad R. *If $MI=M$ then $M=0$.*

Proof. Suppose that M is generated by the set $\{m_1,m_2,\dots,m_s\}$ and that $MI=M$. We proceed by induction on s.

If $s=1$, so M is generated by m_1, then the hypothesis $M=MI$ implies that

$$m_1=m_1'r_1+\cdots+m_t'r_t,$$

for some $m_i'\in M$ and $r_i\in I$. Since $M=m_1R$, it follows that there exists $a_i\in R$ such that $m_i'=m_1a_i$, for each i. Letting $x=a_1r_1+\cdots+a_sr_s\in I$, we get $m_1=m_1x$; whence

$$m_1(1-x)=0. \tag{4.2}$$

Now since $x\in I\subset\operatorname{rad}R$, we see from Lemma 4.1 that $(1-x)$ has a two sided inverse, which we denote by b. Then (4.2) implies $0=m_1(1-x)b=m_1$, and since m_1 generates M, we get $M=0$.

Suppose now that $s\geq 2$. Since $M=MI$, there exist $m\in M$ and $x\in I$ such that $m_1=mx$. On the other hand, M being generated by $m_1,m_2,\dots,m_s$ implies that there exist $a_1,a_2,\dots,a_s$ such that we can write m as $m=m_1a_1+m_2a_2+\cdots+m_sa_s$. Putting these two equations together, we get $m_1=m_1a_1x+m_2a_2x+\cdots+m_sa_sx$; thus

$$m_1(1-a_1x)=m_2a_2x+\cdots+m_sa_sx. \tag{4.3}$$

Now, since $x\in I\subset\operatorname{rad}R$, Lemma 4.1 implies that $(1-a_1x)$ has a two-sided inverse that we denote by b. Then (4.3) implies that $m_1=m_2a_2xb+\cdots m_sa_sxb$, which means that R is already generated by the $s-1$ elements $m_2,\dots,m_s$. By induction, it follows that $M=0$. □

Corollary 4.10. *If A is a finite-dimensional algebra, then* rad A *is nilpotent.*

Proof. First note that, since A is finite-dimensional, then every ideal in A is so too. Therefore, every ideal of A has a finite k-basis and, in particular, every ideal is finitely generated as an A-module. Now, since A is finite-dimensional, the chain of ideals

$$A \supset \operatorname{rad} A \supset (\operatorname{rad} A)^2 \supset (\operatorname{rad} A)^3 \supset \cdots$$

becomes stationary, which means that there exists an integer m such that $(\operatorname{rad} A)^n =$ $\operatorname{rad} A)^m$ for all $n \geq m$. In particular, $(\operatorname{rad} A)^m = (\operatorname{rad} A)^m(\operatorname{rad} A)$, and Lemma 4.9 implies that $(\operatorname{rad} A)^m = 0$. □

We end this section with a standard result about exact sequences of modules.

Lemma 4.11 (Five Lemma). *Given a commutative diagram of R-modules with exact rows*

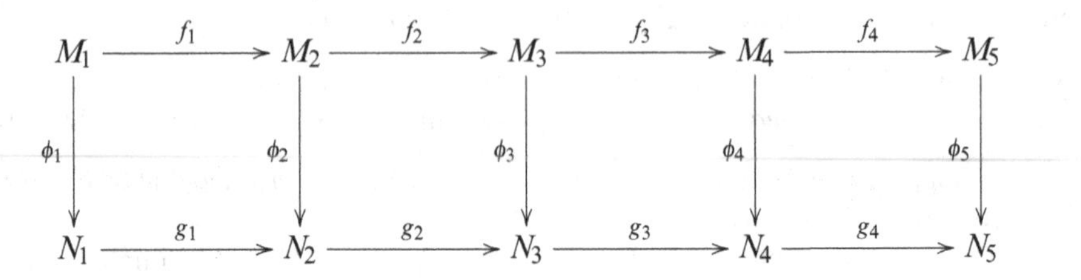

Then

1. *If ϕ_2 and ϕ_4 are surjective and ϕ_5 is injective, then ϕ_3 is surjective.*
2. *If ϕ_1 is surjective and ϕ_2 and ϕ_4 are injective, then ϕ_3 is injective.*
3. *If ϕ_1 is surjective, ϕ_2 and ϕ_4 are isomorphisms, and ϕ_5 is injective, then ϕ_3 is an isomorphism.*

Proof. (3) follows from (1) and (2). We will show (1) and leave (2) as an exercise. Let $n_3 \in N_3$. Since ϕ_4 is surjective, there is $m_4 \in M_4$ such that $\phi_4(m_4) = g_3(n_3)$. By exactness of the second row, we have $g_4 g_3(n_3) = 0$; thus $0 = g_4\phi_4(m_4) = \phi_5 f_4(m_4)$ by commutativity of the diagram. Since ϕ_5 is injective, it follows that $f_4(m_4) = 0$, and then it follows from the exactness of the first row that there exists $m_3 \in M_3$ such that $f_3(m_3) = m_4$. We now have that

$$g_3\phi_3(m_3) = \phi_4 f_3(m_3) = \phi_4(m_4) = g_3(n_3).$$

Thus $g(n_3 - \phi_3(m_3)) = 0$ and the exactness of the second row implies that there exists $n_2 \in N_2$ such that $g_2(n_2) = n_3 - \phi_3(m_3)$. Since ϕ_2 is surjective it follows that there exists $m_2 \in M_2$ such that $\phi_2(m_2) = n_2$; whence $n_3 - \phi_3(m_3) = g_2\phi_2(m_2) = \phi_3 f_2(m_2)$, where the last equation holds because the diagram is commutative. We now have

$$\phi_3(f_2(m_2) + m_3) = \phi_3 f_2(m_2) + \phi_3(m_3) = n_3,$$

which shows that n_3 is in the image of ϕ_3, and thus ϕ_3 is surjective. □

4.4 Idempotents and Direct Sum Decomposition

Let A be a k-algebra.

Definition 4.11. Let $M_1, M_2, \ldots, M_s$ be A-modules. Then the **direct sum** $M_1 \oplus M_2 \oplus \cdots \oplus M_s$ is the A-module whose vector space is the direct sum of the vector spaces of the M_i and whose module structure is given by

$$(m_1, m_2, \ldots, m_s)\, a = (m_1 a, m_2 a, \ldots, m_s a).$$

A module is called **indecomposable** if it cannot be written as the direct sum of two proper submodules.

Recall that the algebra A is itself an A-module; see Example 4.5. The goal of this section is to give a direct sum decomposition of the module A into indecomposable modules. The concept of idempotents is the key to this decomposition.

Definition 4.12. An element $e \in A$ is called an **idempotent** if $e^2 = e$. Two idempotents e_1, e_2 are called **orthogonal** if $e_1 e_2 = e_2 e_1 = 0$. An idempotent e is called central idempotent if $ea = ae$ for all $a \in A$, and a nonzero idempotent is called **primitive** if e cannot be written as $e = e_1 + e_2$ with e_1, e_2 nonzero orthogonal idempotents.

Example 4.8. The elements 0 and 1 are idempotents, called trivial idempotents. All other idempotents are called non-trivial.

Example 4.9. If $A = kQ$ is a path algebra, then each constant path e_i is an idempotent.

Lemma 4.12. *If $A = kQ$ is a path algebra, then each constant path e_i is a primitive idempotent.*

Proof. Suppose that $e_i = e + e'$ with e, e' orthogonal idempotents, say $e = \sum_c \lambda_c c$, $e' = \sum_c \lambda'_c c$, where the sums are over all paths in the quiver.

Since $e = \sum_c \lambda_c c$ is an idempotent, we have

$$0 = e^2 - e = \sum_{c,c'} \lambda_c \lambda_{c'} cc' - \sum_{c''} \lambda_{c''} c'' = \sum_{cc'=c''} (\lambda_c \lambda_{c'} - \lambda_{c''}) c''. \tag{4.4}$$

In particular, $\lambda_{e_j} \lambda_{e_j} = \lambda_{e_j}$ thus $\lambda_{e_j} = 0$ or 1, for any vertex j.

On the other hand, $ee' = 0$ implies $\lambda_{e_j} = 0$, whenever $\lambda'_{e_j} = 1$, and, since $e + e' = e_i$, we must have $\lambda_{e_j} = 0$ and $\lambda'_{e_j} = 0$, if $i \neq j$, and one of $\lambda_{e_i}, \lambda'_{e_i}$ is 0 and the other is 1. Say $\lambda_{e_i} = 1$ and $\lambda'_{e_i} = 0$. Now it follows from (4.4) that for any path c, we have $\lambda_{c'} = 0$; thus $e' = 0$. □

Lemma 4.13. *Let e be a non-trivial idempotent. Then e and $(1 - e)$ are orthogonal idempotents such that $1 = e + (1 - e)$, and the right A-module A is equal to the direct sum*

$$A = eA \oplus (1 - e)A.$$

If in addition e is central, then $A = eA \oplus (1 - e)A$ as k-algebras.

Proof. The computation $(1-e)^2 = 1-e-e+e^2 = 1-e-e+e = 1-e$ shows that $(1-e)$ is an idempotent. Moreover, $e(1-e) = e-e^2 = e-e = 0$ and also $(1-e)e = 0$; thus e and $(1-e)$ are orthogonal idempotents. Now consider the module A. Any element $a \in A$ can be written as $a = ea + a - ea = ea + (1-e)a \in eA + (1-e)A$, thus $A = eA + (1-e)A$. To show that this sum is a direct sum of vector spaces, it is enough to show that the intersection $eA \cap (1-e)A$ contains only 0. So suppose that $a \in eA \cap (1-e)A$. Then $a = ea' = (1-e)a''$, for some $a', a'' \in A$, and therefore $ea' - (1-e)a'' = 0$, whence $0 = e0 = e(ea' - (1-e)a'') = ea'$, where the last equation holds because $e^2 = e$ and $e(1-e) = 0$. This shows that $0 = ea' = a$, and thus $eA \cap (1-e)A = \{0\}$.

This direct sum decomposition respects the right A-module structure of A, since $ab = (ea+(1-e)a)b = eab+(1-e)ab$. If moreover e is central, then $eA\oplus(1-e)A$ is a k-algebra with respect to the componentwise multiplication

$$(ea, (1-e)a) \cdot (ea', (1-e)a') = (eaea', (1-e)a(1-e)a'),$$

since $eaea' = eaa'$ and

$$(1-e)a(1-e)a' = (1-e)aa' - (1-e)eaa' = (1-e)aa'.$$

□

Example 4.10. Let A be the path algebra of the quiver $1 \xrightarrow{\alpha} 2$. Then the constant paths e_1 and $e_2 = 1 - e_1$ are orthogonal idempotents and the right module A decomposes as $e_1A \oplus e_2A$. However e_1 is not central, since $e_1\alpha = \alpha$, whereas $\alpha e_1 = 0$. Thus the direct sum $e_1A \oplus e_2A$ does not reflect the algebra structure of A. Indeed, for any $e_2a \in e_2A$, we have $\alpha(e_2a) \in e_1A$.

By Lemma 4.13, any two orthogonal idempotents whose sum is the identity lead to a direct sum decomposition of the right A-module A. Conversely, it will follow from the next lemma that any direct sum decomposition of the right A-module A produces orthogonal idempotents whose sum is the identity.

Lemma 4.14. *Let $A = M_1 \oplus M_2$ be a direct sum decomposition of the right A-module A. Then*

1. *there exist orthogonal idempotents $e_1 \in M_1$ and $e_2 \in M_2$ such that $1 = e_1 + e_2$,*
2. *M_i is indecomposable if and only if e_i is primitive, for $i = 1, 2$.*

Proof.

(1) Since $A = M_1 \oplus M_2$, every element in A can be written uniquely as a sum of an element in M_1 and an element in M_2. In particular, there exists $e_i \in M_i$ such that $1 = e_1 + e_2$. Since the right A-action on A, and thus

the one on M_i, is given by the multiplication in A, we have that $e_i a \in M_i$ for all $a \in A$. In particular $e_1 e_2 \in M_1$. On the other hand, $1 = e_1 + e_2$ implies $e_2 = e_1 e_2 + e_2^2$ and thus $e_1 e_2 = e_2 - e_2^2 \in M_2$, and therefore $e_1 e_2 \in M_1 \cap M_2$. Since A is the direct sum of M_1 and M_2, we have $M_1 \cap M_2 = \{0\}$, and it follows that $e_1 e_2 = 0$. Similarly, one can show that $e_2 e_1 = 0$, and this proves orthogonality.

Then the equation $e_2 = e_1 e_2 + e_2^2$ implies that $e_2 = e_2^2$, and hence e_2 is an idempotent. In the same way, one can show that e_1 is an idempotent.

(2) ($\Rightarrow$) Suppose that e_1 is not primitive, that is, $e_1 = e + e'$ for two non-trivial orthogonal idempotents $e, e' \in M$. As in the proof of Lemma 4.13, one can show that $M_1 = eM_1 \oplus e'M_1$. Moreover, $eM_1 \neq 0$ because it contains the element $ee_1 = ee + ee' = e \neq 0$; and $e'M_1 \neq 0$ because it contains e'. Thus if e_1 is not primitive, then M_1 is not indecomposable.

($\Leftarrow$) Suppose that M_1 is not indecomposable, that is, $M_1 = M_1' \oplus M_1''$ for some non-zero submodules M_1', M_1'' of M_1. Then we can write $e_1 = e_1' + e_1''$ for some elements $e_1' \in M_1'$ and $e_1'' \in M_1''$. We want to show that e_1', e_1'' are non-trivial orthogonal idempotents. Multiplying the identity

$$1 = e_1 + e_2 = e_1' + e_1'' + e_2$$

with e_1'', we get

$$e_1'' = e_1' e_1'' + e_1'' e_1'' + e_2 e_1'', \tag{4.5}$$

which implies that $e_2 e_1''$ is an element of M_1. But since M_2 is a right A-module and $e_2 \in M_2$, we also have that $e_2 e_1''$ is an element of M_2. Since $M_1 \cap M_2 = \{0\}$, it follows that $e_2 e_1'' = 0$. Thus (4.5) implies that

$$e_1' e_1'' = e_1'' - e_1'' e_1'' \in M_1'',$$

but since M_1' is a right A-module, $e_1' e_1''$ is also in M_1'. Thus $e_1' e_1'' \in M_1' \cap M_1'' = \{0\}$. On the other hand, $e_1 = e_1' + e_1''$ implies that $e_1 = e_1^2 = e_1'^2 + e_1''^2$ with $e_1'^2 \in M_1'$ and $e_1''^2 \in M_1''$, but since $M_1 = M_1' \oplus M_1''$, the decomposition $e_1 = e_1' + e_1''$ is unique, and thus $e_1'^2 = e_1'$ and thus $e_1'^2 = e_1'$ and $e_1''^2 = e_1''$.

To show that e_1' is non-trivial, suppose $e_1' = 0$, let $m' \in M_1', m' \neq 0$ and compute $m' = (e_1 + e_2)m' = e_1 m' + e_2 m'$, with $e_1 m' \in M_1$ and $e_2 m' \in M_2$. But since $M = M_1 \oplus M_2$ and $m' \in M_1$, we must have $m' = e_1 m'$. Therefore

$$m' = e_1 m' = e_1' m' + e_1'' m' = e_1'' m' \in M_1'',$$

hence $m' \in M_1' \cap M_1'' = 0$; whence $m' = 0$, a contradiction.

This shows that $e_1 = e_1' + e_1''$ is the sum of two non-trivial orthogonal idempotents, and therefore e_1 is not primitive. □

Let us assume now that A is finite-dimensional. Then we can write the module A as the direct sum of indecomposable modules:

$$A = M_1 \oplus M_2 \oplus \cdots \oplus M_n.$$

By Lemma 4.14, it follows that there are primitive, pairwise orthogonal idempotents $e_1, e_2, \ldots, e_n$ with $e_i \in M_i$ such that $M_i = e_i A$ and $1 = e_1 + e_2 + \cdots + e_n$.

Conversely, if $e_1, e_2, \ldots, e_n$ are primitive, pairwise orthogonal idempotents such that $1 = e_1 + e_2 + \cdots + e_n$ then, by Lemma 4.13 we have a direct sum decomposition

$$A = e_1 A \oplus e_2 A \oplus \cdots \oplus e_n A,$$

where each $e_i A$ is an indecomposable module.

Example 4.11. Let Q be a quiver without oriented cycles with set of vertices $Q_0 = \{1, 2, \ldots, n\}$, and let $A = kQ$ be its path algebra. By Lemma 4.12, the constant paths e_i are primitive orthogonal idempotents. Moreover $1 = e_1 + e_2 + \cdots + e_n$, and therefore the right A-module A can be written as

$$A = e_1 A \oplus e_2 A \oplus \cdots \oplus e_n A,$$

and each $e_i A$ is an indecomposable module. Note that for each i, the underlying vector space of the module $e_i A$ has basis the set of paths starting at i. Compare $e_i A$ with the projective representation $P(i)$ of the quiver Q.

4.5 A Criterion for Indecomposability

In this section, we show that the question whether a module is indecomposable or not can be answered in terms of its endomorphism algebra. We need the following concept:

Definition 4.13. An algebra A is called **local** if A has a unique maximal right ideal.

Recall that the radical of A is the intersection of all maximal right ideals. So if A is local, then the unique maximal right ideal is equal to the radical of A.

Lemma 4.15. *Let A be a k-algebra. Then the following are equivalent:*

1. *A is local.*
2. *A has a unique maximal left ideal.*
3. *The set of non-invertible elements of A is a two-sided ideal.*
4. *For all $a \in A$, we have a or $(1 - a)$ is invertible.*
5. *The k-algebra $A/\mathrm{rad}\, A$ is a field.*

Proof.

- (1)⇒(3) If A is local, then rad A is the unique maximal right ideal. Since the radical is a proper ideal it cannot contain any invertible elements. On the other hand, if x is not invertible, then $x \in \text{rad } A$, because the right ideal generated by x is not equal to A, and thus it is contained in the unique maximal ideal rad A. This shows that the set of non-invertible elements of A is equal to rad A which is a two-sided ideal by Corollary 4.2.
- (2)⇒(3) is similar and left to the reader.
- (3)⇒(4) If a and $(1 - a)$ were non-invertible then, by (3), we would have that $1 = a + (1 - a)$ is non-invertible, a contradiction.
- (4)⇒(5) We must show that every nonzero element in $A/\text{rad } A$ is invertible, that is, for every $a \in A \setminus \text{rad } A$, there is $c \in A$ such that $(1 - ac) \in \text{rad } A$. Now since $a \notin \text{rad } A$, Lemma 4.1 implies that there exists b such that $(1 - ab)$ has no inverse in A. Then (4) implies that ab has an inverse b'; thus $1 = abb'$. Letting $c = bb'$, the result follows.
- (5)⇒(1) & (2) Condition (5) implies that rad A is a maximal (two-sided) ideal. This shows (1) and (2). □

Corollary 4.16. *If A is a finite-dimensional local k-algebra, then the $A/\text{rad } A$ is isomorphic to the field k.*

Proof. It follows from the lemma that $A/\text{rad } A$ is a field extension of k. Since A is finite-dimensional, this extension is finite-dimensional, and hence it is an algebraic extension. Now the result follows, since k is algebraically closed. □

Remark 4.17. The hypothesis that A is finite-dimensional in the corollary is really necessary. For example, if $A = k(t)$ is the field of rational functions in one variable t, then A is local but not finite-dimensional over k, and $A/\text{rad } A = A = k(t)$ is not isomorphic to k.

The hypothesis that k is algebraically closed is also necessary. For example, if $k = \mathbb{R}$ is the field of real numbers, we can consider $A = \mathbb{C}$, the field of complex numbers, as a k-algebra. Then $A/\text{rad } A = A$ is not isomorphic to k.

Corollary 4.18. *If A is local, then A has only trivial idempotents* 0 *and* 1.

Proof. Suppose that e is an idempotent. Then $e(1 - e) = 0$, by Lemma 4.13. On the other hand, condition (4) of Lemma 4.15 implies that e or $(1 - e)$ is invertible, and thus $e = 0$ or $(1 - e) = 0$. □

Corollary 4.19. *An idempotent $e \in A$ is primitive if and only if the algebra eAe has only trivial idempotents* 0 *and* 1.

Proof. First note that e is the unity element in the algebra eAe.

(⇒) Let e be a primitive idempotent. Let e' be an idempotent in eAe, thus $e' = eae$, for some $a \in A$. Then $(e - e')$ is an idempotent too, and $e'(e - e') = 0$, by Lemma 4.13. Moreover, $e = e' + (e - e')$, and, since e is primitive, we must have $e' = 0$ or $e - e' = 0$. This shows $e - e' = 0$. This shows that e' is a trivial idempotent; hence all idempotents of eAe are trivial.

($\Leftarrow$) Suppose that $e = e' + e''$ where e', e'' are orthogonal idempotents in A. Then

$$(ee'e)(ee'e) = ee'ee'e = ee'(e' + e'')e'e = ee'e + ee'e''e'e = ee'e$$

where the last equation follows from $e'e'' = 0$. Thus $(ee'e)$ is an idempotent in A and hence in eAe. By our hypothesis, it follows that $ee'e = 0$ or $ee'e = e$. In the first case, we have $0 = ee'e = (e'+e'')e'(e'+e'') = e'$ since $e'e'' = e''e' = 0$, and in the second case, we have $e' + e'' = e = ee'e = (e' + e'')e'(e' + e'') = e'$; thus $e'' = 0$. This shows that e cannot be written as the sum of two non-trivial orthogonal idempotents; hence e is primitive. □

Corollary 4.20. *Let A be a k-algebra, let M be a finite-dimensional A-module, and let* $\operatorname{End} M$ *be its endomorphism algebra. Then the following are equivalent:*

1. *M is indecomposable.*
2. *Every endomorphism $f \in \operatorname{End} M$ is of the form $f = \lambda 1_M + g$, with $g \in \operatorname{End} M$ nilpotent and $\lambda \in k$.*
3. $\operatorname{End} M$ *is local.*

Proof.

(1 $\Rightarrow$ 2) Let f be an endomorphism of M. So, in particular, $f: M \to M$ is a k-linear map between finite-dimensional k-vector spaces. Since the field k is algebraically closed, the characteristic polynomial of f can be written as

$$\chi_f(x) = \prod_{i=1}^{t} (x - \lambda_i)^{\nu_i}.$$

Thus λ_i are the eigenvalues of f, and there is a basis B of M such that f is represented by a triangular matrix whose diagonal entries are the eigenvalues λ_i with multiplicities ν_i. Let $M_i = \ker(f - \lambda_i 1_M)^{\nu_i}$. Then $\dim M_i = \nu_i$, and the vector space M decomposes as

$$M = M_1 \oplus \cdots \oplus M_t. \tag{4.6}$$

Now let $h_i = (f - \lambda_i 1_M)^{\nu_i}$. Then h_i is a polynomial in f, that is,

$$h_i = f^{\nu_i} + a_{\nu_i - 1} f^{\nu_i - 1} + \cdots + a_1 f + a_0 1_M,$$

for some $a_j \in k$. And since $f \in \operatorname{End} M$, we also have $h_i \in \operatorname{End} M$, and therefore its kernel $M_i = \ker h_i$ is an A-module. This shows that (4.6) is actually a direct sum decomposition of M into A-modules. Since M is indecomposable, it follows that $t = 1$, and so f has only one eigenvalue λ. Therefore the matrix of f with respect to the basis B is a triangular matrix whose diagonal entries are all equal to λ, and thus $f = \lambda 1_M + g$ with g nilpotent.

(2 ⇒ 3) Let $f = \lambda 1_M + g \in \operatorname{End} M$. If f is not invertible, then $\lambda = 0$ and $f = g$ is nilpotent. Thus there exists $\ell \geq 0$ such that $f^\ell = 0$. But then

$$1_M = 1_M - f^\ell = (1_M + f + f^2 + \cdots + f^{\ell-1})(1 - f),$$

which shows that $(1 - f)$ is invertible, and then Lemma 4.15 implies that End M is local.

(3 ⇒ 1) Assume that End M is local and suppose that $M = M_1 \oplus M_2$. Let $p_i : M \to M_i$ be the canonical projection, and let $u_i : M_i \to M$ be the canonical injection, for $i = 1, 2$. Then $u_i \circ p_i \in \operatorname{End} M$ and $(u_i \circ p_i)^2 = u_i \circ p_i$. Thus $u_i \circ p_i$ is an idempotent in End M, which implies that $u_i \circ p_i = 0$ or 1, since End M is local. If $u_i \circ p_i = 0$, then $M_i = 0$ and if $u_i \circ p_i = 1$, then $M_i = M$. This shows that M is indecomposable.

□

We end this chapter with two examples of endomorphism algebras of indecomposable modules over path algebras. We represent the modules as representations, using the fact that the category of finitely generated modules over a path algebra of a quiver Q is equivalent to the category of finite-dimensional representations of Q, see Theorem 5.4.

Example 4.12. If A is the path algebra of a quiver of Dynkin type $\mathbb{A}$, $\mathbb{D}$ or $\mathbb{E}$, then it follows from our computation of the dimension of the Hom spaces in Chap. 3, that End $M = k$ for any indecomposable representation M of this type. Clearly, the field k is local.

On the other hand, if M is not indecomposable, then the identity morphism on each indecomposable summand of M is a non-trivial idempotent in End M; thus End M is not a local algebra.

Example 4.13. Let Q be the quiver $1 \overset{\alpha}{\underset{\beta}{\rightrightarrows}} 2$, and let M be the representation below:

$$k^2 \overset{\begin{bmatrix} 1 & 0 \\ 0 & 1 \end{bmatrix}}{\underset{\begin{bmatrix} 1 & \lambda \\ 0 & 1 \end{bmatrix}}{\rightrightarrows}} k^2$$

Then an endomorphism of M is a morphism of representations $f : M \to M$, so it consists of two linear maps $f_1, f_2 : k^2 \to k^2$ which commute with the two maps $\varphi_\alpha, \varphi_\beta$ of the representation M. Since φ_α is given by the identity matrix, we see that f_1 and f_2 must have the same matrix. Let us denote this matrix by

$$\begin{bmatrix} a & b \\ c & d \end{bmatrix}.$$

Moreover, since $f_2\varphi_\beta = \varphi_\beta f_1$, it follows that

$$\begin{bmatrix} a & \lambda a + b \\ c & \lambda c + d \end{bmatrix} = \begin{bmatrix} a + \lambda c & b + \lambda d \\ c & d \end{bmatrix}.$$

If $\lambda \neq 0$, this equation implies that $c = 0$ and $a = d$. Thus End M is the algebra of all matrices of the form

$$\begin{bmatrix} a & b \\ 0 & a \end{bmatrix}.$$

Note that an element of End M is invertible if and only if $a \neq 0$. On the other hand, if $a = 0$, then

$$\begin{bmatrix} 1 & 0 \\ 0 & 1 \end{bmatrix} - \begin{bmatrix} a & b \\ 0 & a \end{bmatrix}$$

is invertible. It follows from Lemma 4.15 that End M is local, and from Corollary 4.20 we get that M is indecomposable.

On the other hand, if $\lambda = 0$, then End M is isomorphic to the algebra of all 2×2 matrices. This algebra is not local, since

$$\begin{bmatrix} 1 & 0 \\ 0 & 0 \end{bmatrix} \quad \text{and} \quad \begin{bmatrix} 1 & 0 \\ 0 & 1 \end{bmatrix} - \begin{bmatrix} 1 & 0 \\ 0 & 0 \end{bmatrix}$$

are both not invertible.

4.6 Notes

For further information on Ring Theory, we refer to [1, 31, 45–47] and for information on algebras to [8, 15, 49, 52, 55, 56].

Problems

Exercises for Chap. 4

4.1. Let A be a k-algebra and let M be a right A-module. Show that M is also a k-vector space.

4.2. Check that the module axioms are satisfied in the numbers 4 and 5 of Example 4.5.

4.3. Let $0 \longrightarrow L \xrightarrow{f} M \xrightarrow{g} N \longrightarrow 0$ be a short exact sequence of A-modules. Show the following:

1. If X is an A-module and $h : X \to M$ is a morphism such that $gh = 0$, then there is a morphism $h' : X \to L$ such that $fh' = h$.
2. If X is an A-module and $h : M \to X$ is a morphism such that $hf = 0$, then there is a morphism $h' : N \to X$ such that $gh' = h$.

4.4. Prove Proposition 4.8

4.5. Let G be a group and let

$$kG = \left\{ \sum_{g \in G} \lambda_g g \mid \lambda_g \in k, \text{ only finitely many } \lambda_g \text{ nonzero} \right\}$$

be the k-algebra with basis G and multiplication given by the group operation. kG is called the *group algebra* of G. Show that

1. $k\mathbb{Z}$ is isomorphic to the algebra of Laurent polynomials in one variable.
2. $k\,(\mathbb{Z}/n\mathbb{Z})$ is isomorphic to $k[x]/(x^n - 1)$.

4.6. Let G be a cyclic group of order n with generator σ, and let kG be the group algebra as in Exercise 4.5.

1. Show that $e_1 = \frac{1}{n}\sum_{g \in G} g$ is an idempotent in kG.
2. Let $\omega \in k$ be a primitive nth root of unity, that is, $\omega^n = 1$ and $\omega^m \neq 1$ for $m = 1, 2, \ldots, n-1$. Show that $e_j = \frac{1}{n}\sum_{i=1}^{n}(w^j\sigma)^i$ is an idempotent in kG.
3. Show that $e_i e_j = 0$ if $i \neq j$ and that $e_1 + e_2 + \cdots + e_n = 1_{kG}$. [Hint: Use that the sum of all nth roots of unity is equal to 0.]

4.7. Let M be an A-module and $X \subset M$ a subset of M. The *annihilator* Ann X of X is defined as

$$\operatorname{Ann} X = \{a \in A \mid xa = 0 \text{ for all } x \in X\}.$$

The module M is called *faithful* if $\operatorname{Ann} M = 0$. Prove that

1. $\operatorname{Ann} X$ is a right ideal in A.
2. If X is a submodule of M, then $\operatorname{Ann} X$ is a two-sided ideal.
3. M is an $A/\operatorname{Ann} M$-module.
4. M is faithful over $A/\operatorname{Ann} M$.

4.8. Let M be an A-module and $\operatorname{Ann} M$ its annihilator as in Exercise 4.7. For any ideal I in A, show that M is an A/I-module by the rule $m(a + I) = ma$ if and only if $I \subset \operatorname{Ann}(M)$.

4.9. Let $0 \to K \to P \xrightarrow{\varphi} M \to 0$ and $0 \to K' \to P' \xrightarrow{\varphi'} M \to 0$ be two short exact sequences of R-modules where P and P' are projective. Show that $P \oplus K' \cong P' \oplus K$.

4.10. Let $0 \to L \to M \to N \to 0$ be a short exact sequence of R-modules, and let $P_1 \to P_0 \to L \to 0$ and $P_1' \to P_0' \to N \to 0$ be two short exact sequences of R-modules with P_0, P_1, P_0', P_1' projective. Show that there is a commutative diagram of the form

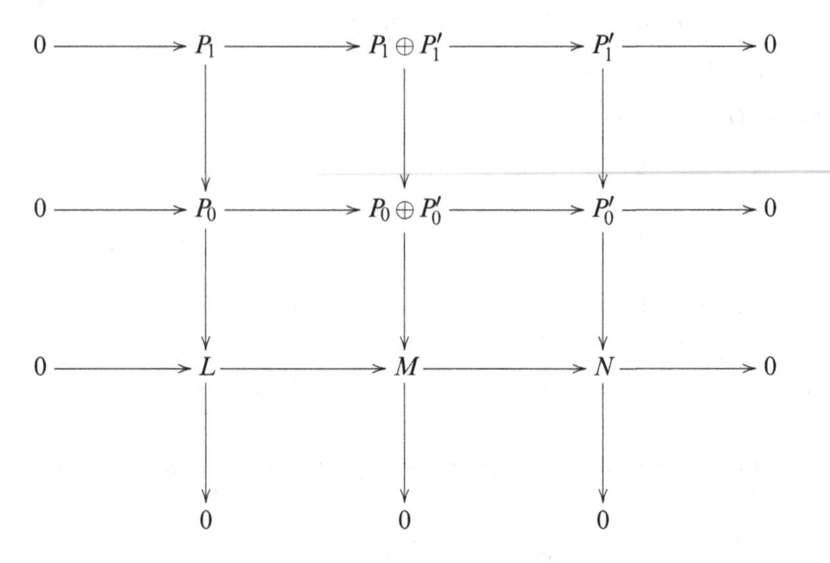

whose rows and columns are exact.

4.11. Let Q be the quiver $1 \overset{\alpha}{\underset{\beta}{\rightrightarrows}} 2$, and let M be the representation below:

$$k^2 \overset{\begin{bmatrix} 1 & 0 \\ 0 & 1 \\ 0 & 0 \end{bmatrix}}{\underset{\begin{bmatrix} 0 & 0 \\ 1 & 0 \\ 0 & 1 \end{bmatrix}}{\rightrightarrows}} k^3$$

1. Compute End M.
2. Show that M is indecomposable.

4.12. Let Q be the quiver

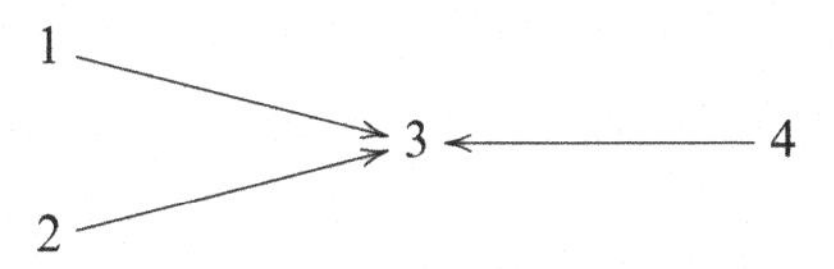

and let M be the representation below:

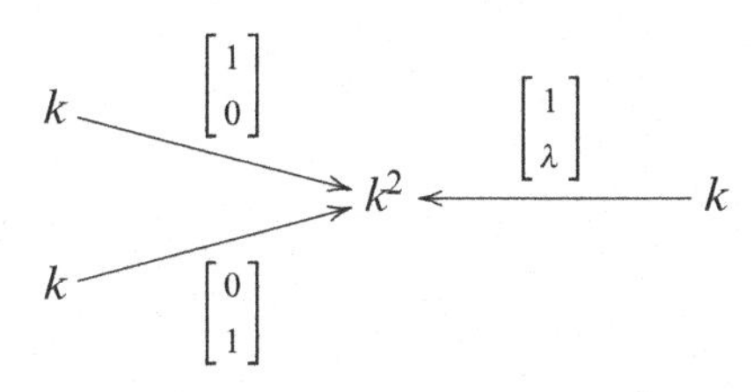

1. Compute End M.
2. Show that End M is local if and only if $\lambda \neq 0$. Hence M is indecomposable if and only if $\lambda \neq 0$.

Chapter 5
Bound Quiver Algebras

A bound quiver algebra is the quotient of a path algebra kQ by an ideal I which is required to satisfy a certain admissibility condition. Bound quiver algebras play a central role in representation theory, since, for any finite-dimensional algebra A over an algebraically closed field k, the category mod A is equivalent to the category mod kQ/I, for some bound quiver algebra kQ/I.

The main result in this chapter is Theorem 5.4 which states that the category mod kQ/I is equivalent to the category rep (Q, I) of all those representations of the quiver Q which satisfy the relations induced by the ideal I. Summarizing we may say that understanding finitely generated modules over a k-algebra is equivalent to understanding finite-dimensional quiver representations.

In the first section, we introduce admissible ideals and bound quiver algebras, and we prove Theorem 5.4 in the second section. In the third section, we use the equivalence of categories of Theorem 5.4 to describe the indecomposable projective and the indecomposable injective representations of bound quiver algebras. We then introduce the notions of projective dimension and global dimension in the fourth section and give examples of Auslander–Reiten quivers of bound quiver algebras in the last section.

5.1 Admissible Ideals and Quotients of Path Algebras

Let $Q = (Q_0, Q_1)$ be a finite quiver, and let $A = kQ$ be its path algebra. If the quiver Q has oriented cycles, then the path algebra is infinite-dimensional. We want to consider quotients of the path algebra by certain ideals, and we would like these quotients to be finite-dimensional and indecomposable as an algebra. This leads to the concept of admissible ideals.

R. Schiffler, *Quiver Representations*, CMS Books in Mathematics,
DOI 10.1007/978-3-319-09204-1_5

First we define the **arrow ideal** R_Q of A to be the two-sided ideal generated by all arrows in Q. As a vector space, we can decompose the arrow ideal as

$$R_Q = kQ_1 \oplus kQ_2 \oplus \cdots \oplus kQ_\ell \oplus \cdots$$

where kQ_ℓ is the subspace of kQ with basis the set Q_ℓ of paths of length ℓ. The ℓ-th power of the arrow ideal can be decomposed as

$$R_Q^\ell = \bigoplus_{m \geq \ell} kQ_m,$$

and it has a basis consisting of all paths of length greater or equal to ℓ.

Definition 5.1. A two-sided ideal I of kQ is called an **admissible ideal** if there exists an integer $m \geq 2$ such that

$$R_Q^m \subset I \subset R_Q^2.$$

If I is an admissible ideal of kQ, then (Q, I) is called a **bound quiver** and the quotient algebra kQ/I is called a **bound quiver algebra**.

Remark 5.1. The condition that $R_Q^m \subset I$ implies that the admissible ideal I contains all paths of length greater or equal to m, which guarantees that the bound quiver algebra is finite-dimensional.

If the quiver Q does not contain any oriented cycles, then there always exists m such that R_Q^m is a subset of I, it suffices to take m greater then the length of the longest path in Q. Thus if Q has no oriented cycles, an ideal is admissible if and only if it is contained in R_Q^2.

The condition $I \subset R_Q^2$ guarantees that we do not cut any arrows when we take the quotient; thus the bound quiver algebra is connected.

Remark 5.2. Suppose that I is an admissible ideal which is generated by the elements $\sigma_1, \sigma_2, \ldots, \sigma_s$. For every pair of vertices x, y, the element $e_x \sigma_i e_y$ is a linear combination of paths from x to y, hence a relation. Since $\sigma_i = \sum_{x,y} e_x \sigma_i e_y$, we see that the ideal I is also generated by the set of relations $\{e_x \sigma_i e_y \mid i = 1, 2, \ldots, s;\ x, y \in Q_0\}$. This shows that for every admissible ideal I, there exists a set of relations that generate I.

Remark 5.3. A finite-dimensional algebra A is called *basic* if for every set of primitive orthogonal idempotents $\{e_1, \ldots, e_n\}$ such that $1 = e_1 + \cdots + e_n$, we have

$$e_i A \cong e_j A \Longleftrightarrow i = j.$$

If A is not basic, define $e_A = e_{s_1} + \cdots e_{s_t}$ to be the sum of a maximal set of primitive orthogonal idempotents such that $e_{s_i} A \cong e_{s_j} A \Leftrightarrow i = j$. Then it can be shown that the algebra $e_A A e_A$ is basic and that the module categories of A and $e_A A e_A$ are equivalent [8, I.6]. Hence, from the point of view of representation theory, it suffices to consider only basic algebras.

It can also be shown that every basic finite-dimensional k-algebra is isomorphic to a quotient of a path algebra by an admissible ideal [8, II.4]. The vertices of the quiver of that path algebra are in bijection with a set of primitive, orthogonal idempotents $\{e_1, e_2, \ldots, e_n\}$ with the property that $1 = e_1 + \cdots + e_n$, and the number of arrows from e_i to e_j is the dimension of the vector space $e_i \left(\operatorname{rad} A/(\operatorname{rad} A)^2\right) e_j$.

We can rephrase Remark 5.3 as follows:

> From the point of view of representation Theory, the study of finite-dimensional k-algebras reduces to the study of bound quiver algebras.

Example 5.1. Let Q be the quiver

$$\alpha \circlearrowleft 1 \xrightarrow{\ \beta\ } 2\,.$$

Then the ideal $I = \langle \alpha^2\beta, \alpha^3 \rangle$ is admissible.

Indeed, we can take $m = 3$; then any path of length greater or equal to 3 must contain α^3 or $\alpha^2\beta$ as a subpath. Thus $R_Q^3 \subset I$. On the other hand, it is clear that $I \subset R_Q^2$ since the generators of I are of length 3.

The following example shows that different relations on the same quiver may lead to isomorphic algebras. Further examples of this fact are given in the exercises.

Example 5.2. Let Q be the quiver

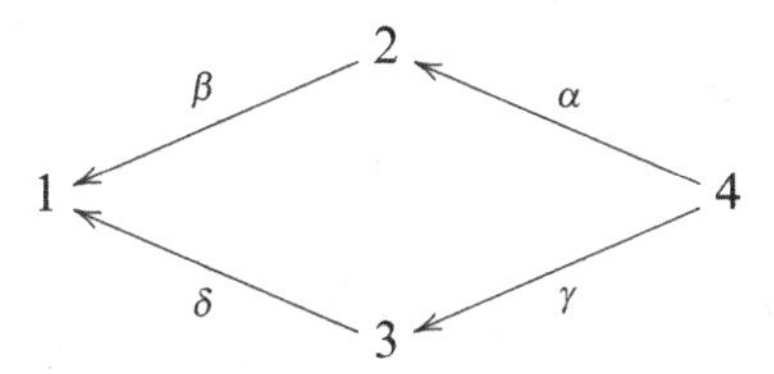

and define two admissible ideals $I_1 = \langle \alpha\beta + \gamma\delta \rangle$ and $I_2 = \langle \alpha\beta - \gamma\delta \rangle$. Then $I_1 \neq I_2$ (unless the characteristic of k is 2), but the corresponding bound quiver algebras are isomorphic:

$$kQ/I_1 \cong kQ/I_2.$$

5.2 Equivalence of the Categories rep (Q, I) and mod kQ/I

Starting from a quiver Q, we have studied two categories: the category of finite-dimensional quiver representations rep Q and the category of finitely generated kQ-modules mod kQ. In this section, we will prove that the two categories are equivalent. Actually, we show a more general result, namely that for any admissible ideal I the category rep (Q, I) of finite-dimensional bound quiver representations is equivalent to the category mod kQ/I of finitely generated kQ/I-modules.

This shows that modules and representations are essentially the same. It is very convenient for us to be able to switch between the two descriptions and to use the one that is best suited to our needs in the given circumstances. For instance, when computing Auslander–Reiten quivers, it is more convenient to use the graphical notation coming from the representations of the quiver. On the other hand, in algebraic arguments, it is often easier to use modules instead of representations.

Theorem 5.4. *Let $A = kQ/I$, where Q is a finite connected quiver and I is an admissible ideal. Then there is an equivalence of categories between the category* mod A *of finitely generated right A-modules and the category* rep (Q, I) *of finite-dimensional bound quiver representations:*

$$\operatorname{mod} A \cong \operatorname{rep}(Q, I).$$

Proof. We will explicitly construct two functors

$$F : \operatorname{mod} A \to \operatorname{rep}(Q, I) \quad \text{and} \quad G : \operatorname{rep}(Q, I) \to \operatorname{mod} A$$

such that $F \circ G \cong 1_{\operatorname{rep}(Q,I)}$ and $G \circ F \cong 1_{\operatorname{mod} A}$.

Let us start with F. We have to define F on A-modules and on morphisms. Let M be an A-module. Then $F(M)$ is the representation $(M_i, \varphi_\alpha)_{i \in Q_0, \alpha \in Q_1}$, where $M_i = Me_i$ is the vector space consisting of all me_i with $m \in M$, and for any arrow $i \xrightarrow{\alpha} j$, the map $\varphi_\alpha : M_i \to M_j$ is given by

$$\varphi_\alpha(me_i) = m(e_i\alpha) = \begin{cases} m\alpha \text{ if } s(\alpha) = i; \\ 0 \quad \text{otherwise.} \end{cases}$$

Let $f : M \to M'$ be a morphism in mod A. Then its image under F is the morphism of quiver representations defined on the vertex i as the map $f_i : M_i \to M'_i$ sending me_i to $f(m)e_i$.

<u>F is well defined:</u> Each Me_i is a vector space and each φ_α is a linear map, so $F(M)$ is a representation of Q. To show that $F(f)$ is a morphism of representations, let $i \xrightarrow{\alpha} j$ be an arrow in Q. Then $f_j \circ \varphi_\alpha(me_i) = f(m)(e_i\alpha) = \varphi'_\alpha \circ f_i(me_i)$, which shows that the diagram

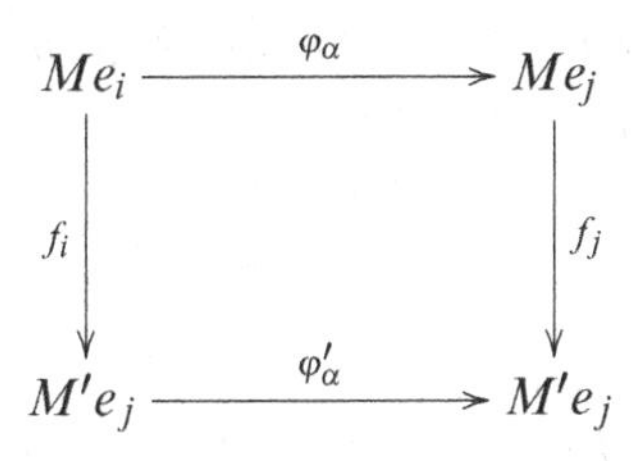

commutes. Finally, we must show that the representation $F(M)$ satisfies the relations in I. Let $\rho = \sum \lambda_c c$ be a relation in I. We must show that $\sum \lambda_c \varphi_c = 0$, where, to each path $c = (i \mid \alpha_1, \alpha_2, \ldots, \alpha_t \mid j)$, the map φ_c is the composition $\varphi_{\alpha_t} \circ \cdots \circ \varphi_{\alpha_1}$. Thus for me_i in M_i, we have $\varphi_c(me_i) = me_i\alpha_1\alpha_2\cdots\alpha_t = mc$, and therefore

$$\sum \lambda_c \varphi_c(me_i) = \sum \lambda_c mc = m \sum \lambda_c c = m\rho.$$

But $\rho \in I$, thus $\rho = 0$ in A, and therefore $m\rho = 0$, as required. This shows that $F(M)$ is an object in rep (Q, I).

It is clear that $F(1_X) = 1_{F(X)}$ for every A-module X and that $F(fg) = F(f)F(g)$, for any two morphisms $f : X \to Y$ and $g : Y \to Z$ in mod A. Thus F is a functor.

Now we define the functor G. We want to associate an A-module M to a quiver representation $(M_i, \varphi_\alpha)_{i \in Q_0, \alpha \in Q_1}$. The underlying vector space of M is the direct sum $M = \bigoplus_{i \in Q_0} M_i$, on which the right A-action is defined as follows: Let $m = (m_i)_{i \in Q_0} \in M$, and let a be an element in A, say $a = \sum \lambda_c c + I$, where the sum is over all paths c in the quiver Q. Define

$$m \cdot a = \sum \lambda_c \varphi_c(m), \tag{5.1}$$

where $\varphi_c(m) = (0, \ldots, 0, \varphi_c(m_{s(c)}), 0, \ldots, 0)$ with the unique nonzero entry at position $t(c)$.

To show that this A-action is well defined, we must show that the right-hand side of equation (5.1) does not depend on the choice of the representative $\sum \lambda_c c$. Suppose that $\sum \mu_c c$ is another representative of the coset a; thus $\sum \lambda_c c - \sum \mu_c c \in I$. Then $\sum \lambda_c \varphi_c(m) - \sum \mu_c \varphi_c(m) = \sum (\lambda_c - \mu_c)\varphi_c(m)$. But since (M_i, φ_α) is a representation of the bound quiver (Q, I), the map $\sum(\lambda_c - \mu_c)\varphi_c$ is the zero map. Thus the right A-action in equation (5.1) is well defined.

So far we have defined G on the objects of rep (Q, I). On morphisms $f = (f_i)_{i \in Q_0} : (M_i, \varphi_\alpha) \to (M'_i, \varphi'_\alpha)$ in rep (Q, I), define $G(f) : M \to M'$ by

$$G(f)(m) = (f_i(m_i))_{i \in Q_0}.$$

G is well defined: First we show that M is an A-module. Since it is clearly a k-vector space, it only remains to show that the four axioms of Definition 4.8 hold. Let $m, m' \in M$, $\lambda \in k$, and $a, a' \in A$. It suffices to show the axioms for the special case where $a = c + I$, $a' = c' + I$ are represented by paths c, c'. Then

(1) $(m + m')a = \varphi_c(m + m') = \varphi_c(m) + \varphi_c(m') = ma + m'a$.
(2) $m(a + a') = (\varphi_c + \varphi_{c'})(m) = \varphi_c(m) + \varphi_{c'}(m) = ma + ma'$.
(3) $m(aa') = \varphi_{cc'}(m) = \varphi_{c'} \circ \varphi_c(m) = (ma)a'$.
(4) $m1 = m \sum_{i \in Q_0} e_i = \sum_{i \in Q_0} \varphi_{e_i}(m) = \sum_{i \in Q_0} m_i = m$.

This shows that M is an A-module.

In order to show that $G(f) : M \to M'$ is a morphism, let $m = (m_i)_{i \in Q_0} \in M$ and $a = c + I \in A$. Since each f_i is linear, the map $G(f)$ is also linear. Moreover $G(f)(ma) = (f_i(\varphi_c(m))_i$ which is equal to $f_{t(c)}(\varphi_c(m_{s(c)}))$ at position $t(c)$; and zero elsewhere. On the other hand, $G(f)(m)a = (f_i(m_i))_i\, a$ which is equal to $\varphi'_c(f_{s(c)}(m_{s(c)}))$ at position $t(c)$; and zero elsewhere. Since f is a morphism of quiver representations, it follows that $\varphi'_c \circ f_{s(c)} = f_{t(c)} \circ \varphi_c$, for every path c. Thus $G(f)(ma) = G(f)(m)a$, and $G(f)$ is a morphism in mod A.

Clearly $G(1_X) = 1_{G(X)}$ and $G(fg) = G(f)G(g)$ for all morphisms $f, g \in$ rep (Q, I). Thus G is a functor.

It remains to prove that $F \circ G \cong 1_{\text{rep}\,(Q,I)}$ and $G \circ F \cong 1_{\text{mod}\,A}$.

Let $(M_i, \varphi_\alpha)_{i \in Q_0, \alpha \in Q_1} \in$ rep (Q, I) and denote its image under G simply by M. Then the representation $F(M)$ at the vertex i is just $Me_i = (\oplus M_i)e_i = M_i$, and the linear map $M_i \to M_j$ on an arrow $i \xrightarrow{\alpha} j$ in $F(M)$ maps m_i to $m_i\alpha = \varphi_\alpha(m_i)$. Thus $F(M) = (M_i, \varphi_\alpha)$, and $F \circ G \cong 1_{\text{rep}\,(Q,I)}$.

On the other hand, let $M \in$ mod A and denote $F(M)$ by (M_i, φ_α). Then the underlying vector space of the module $G \circ F(M)$ is $\oplus M_i = \oplus Me_i \cong M$, and the A-action on it is given by $m \cdot \sum \lambda_c c = \sum \lambda_c \varphi_c(m) = \sum \lambda_c mc$. Thus $G \circ F \cong 1_{\text{mod}\,A}$, and this proves that F and G are equivalences of categories. □

Remark 5.5. Many of the results on quiver representations which we have proved in Chap. 2 also hold for bound quiver representations and modules. Examples are Theorem 2.11 and Proposition 2.29. We leave the proofs as an exercise.

Definition 5.2. An algebra A is called **hereditary** if each submodule of a projective module is projective.

Proposition 5.6. *Path algebras of quivers without oriented cycles are hereditary.*

Proof. This follows from the equivalence of categories in Theorem 5.4 and the corresponding result on quiver representations in Theorem 2.24. □

5.3 Projective Representations of Bound Quivers

In Sect. 2.1, we have defined the indecomposable projective representations $P(i)$, as well as the indecomposable injective representations $I(i)$ for a quiver without relations. Now we do the same for bound quivers.

Let (Q, I) be a bound quiver and $A = kQ/I$ its bound quiver algebra. For every vertex $i \in Q_0$, we will define an indecomposable projective representation $P(i)$ and an indecomposable injective representation $I(i)$ of the bound quiver (Q, I).

Definition 5.3. Let i be any vertex in Q.

(a)

$$P(i) = (P(i)_j, \varphi_\alpha)_{j\in Q_0, \alpha\in Q_1}$$

where $P(i)_j$ is the k-vector space with basis the set of all residue classes $c + I$ of paths c from i to j in Q;
and if $j \xrightarrow{\alpha} \ell$ is an arrow in Q, then $\varphi_\alpha : P(i)_j \to P(i)_\ell$ is the linear map defined on the basis by composing the paths from i to j with the arrow $j \xrightarrow{\alpha} \ell$, that is,

$$\varphi_\alpha(c + I) = c\alpha + I.$$

(b)

$$I(i) = (I(i)_j, \varphi_\alpha)_{j\in Q_0, \alpha\in Q_1}$$

where $I(i)_j$ is the k-vector space with basis the set of all residue classes $c + I$ of paths c from j to i in Q;
and if $j \xrightarrow{\alpha} \ell$ is an arrow in Q, then $\varphi_\alpha : I(i)_j \to I(i)_\ell$ is the linear map defined on the basis by deleting the arrow $j \xrightarrow{\alpha} \ell$ from those paths from j to i which start with α and sending to zero the paths that do not start with α, that is,

$$\varphi_\alpha(c + I) = \begin{cases} c' + I & \text{if } c = \alpha c', \\ 0 & \text{otherwise.} \end{cases}$$

Proposition 5.7. *Under the equivalence of categories of Theorem 5.4, $P(i)$ corresponds to the indecomposable projective module $e_i A$, and $I(i)$ corresponds to the indecomposable injective module DAe_i.*

Proof. Consider the equivalence $G\colon \text{rep}(Q, I) \to \text{mod}\, A$ from Theorem 5.4. Then $G(P(i))$ is the module whose underlying vector space is the direct sum $\oplus_{j\in Q_0} P(i)_j$. This vector space has a basis consisting of all residue classes $c + I$ of paths c starting at i and is therefore isomorphic to the vector space $e_i A$. It remains to check that the right A-action on $G(P(i))$ coincides with the one on $e_i A$. A residue class $\tilde{c} + I \in A$ of a path $\tilde{c}$ acts on a basis element $c + I$ of $G(P(i))$ by the formula $(c + I) \cdot (\tilde{c} + I) = \varphi_{\tilde{c}}(c)$, which is given by the composition of paths $c\tilde{c}$. This shows that $G(P(i)) = e_i A$. The second statement of the proposition is proved in a similar way. □

Example 5.3. Let Q be the quiver

$$1 \xrightarrow{\alpha} 2 \xrightarrow{\beta} 3 \xrightarrow{\gamma} 4$$

and $I = \langle \alpha\beta\gamma \rangle$. Then

$$P(1) = \begin{smallmatrix} 1 \\ 2 \\ 3 \end{smallmatrix}, \; P(2) = \begin{smallmatrix} 2 \\ 3 \\ 4 \end{smallmatrix}, \; P(3) = \begin{smallmatrix} 3 \\ 4 \end{smallmatrix}, \; P(4) = 4 .$$

Example 5.4. Let Q be the quiver

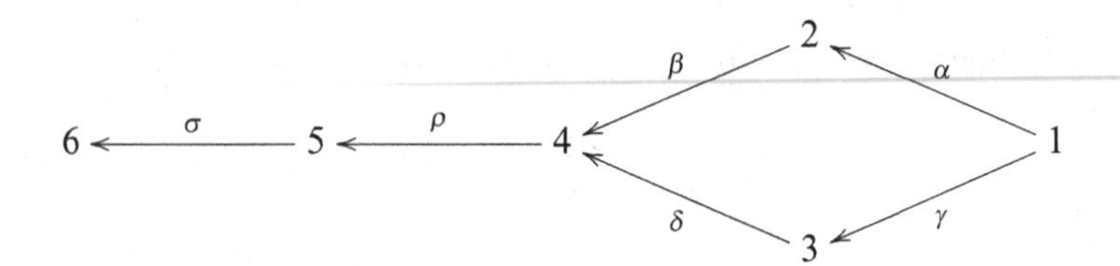

and $I = \langle \alpha\beta - \gamma\delta,\ \beta\rho,\ \delta\rho\sigma \rangle$. Then

$$P(1) = \begin{smallmatrix} & 1 & \\ 2 & & 3 \\ & 4 & \end{smallmatrix}, \; P(2) = \begin{smallmatrix} 2 \\ 4 \end{smallmatrix}, \; P(3) = \begin{smallmatrix} 3 \\ 4 \\ 5 \end{smallmatrix},$$

$$P(4) = \begin{smallmatrix} 4 \\ 5 \\ 6 \end{smallmatrix}, \; P(5) = \begin{smallmatrix} 5 \\ 6 \end{smallmatrix}, \; P(6) = 6 .$$

Example 5.5. Let Q be the quiver

$$\alpha \circlearrowright 1 \xrightarrow{\beta} 2$$

and $I = \langle \alpha^2\beta, \alpha^3 \rangle$. Then $P(2)$ is the simple representation $S(2)$ and $P(1)_1$ is the three-dimensional vector space with ordered basis $\{e_1, \alpha, \alpha^2\}$ and $P(1)_2$ is the two dimensional-vector space with ordered basis $\{\beta, \alpha\beta\}$, and the linear maps are given with respect to these bases by the matrices:

$$\varphi_\alpha = \begin{bmatrix} 0 & 0 & 0 \\ 1 & 0 & 0 \\ 0 & 1 & 0 \end{bmatrix} \text{ and } \varphi_\beta = \begin{bmatrix} 1 & 0 & 0 \\ 0 & 1 & 0 \end{bmatrix}.$$

Next we define the radical of a module.

Definition 5.4. (a) Let A be a finite-dimensional k-algebra, and let $M \in \operatorname{mod} A$. The intersection of all maximal submodules of M is called the **radical** $\operatorname{rad} M$ of the module M.
(b) The quotient $M/\operatorname{rad} M$ is called the **top** of M and is denoted by $\operatorname{top} M$.

Remark 5.8. One can show that $\operatorname{top} M$ is a semisimple module, which means that it is a direct sum of simple modules.

Lemma 5.9. *Let $A = kQ/I$ be a bound quiver algebra and $P(i) = (P(i)_j, \varphi_\alpha)$ the indecomposable projective representation at vertex i. Then*

$$\operatorname{rad} P(i) = (P(i)'_j, \varphi'_\alpha),$$

where $P(i)'_j = P(i)_j$ if $i \neq j$, and $P'(i)_i$ is the vector space spanned by

$$\{c + I \mid c \text{ is a nonconstant path from } i \text{ to } i\};$$

and

$$\varphi'_\alpha = \varphi_\alpha|_{P(i)'_{s(\alpha)}}.$$

In particular, $\operatorname{top} P(i) \cong S(i)$.

Proof. Let $M = (P(i)'_j, \varphi'_\alpha)$ be as in the lemma. Let $f = (f_j)_{j \in Q_0} : M \to P(i)$ be the inclusion map; thus $f_j(c + I) = c + I$ for all $j \in Q_0$ and all path c from i to j. To see that f is a morphism of quiver representations, it suffices to check that for every arrow α we have a commutative diagram:

$$\begin{array}{ccc} P(i)'_{s(\alpha)} & \xrightarrow{\varphi'_\alpha} & P(i)'_{t(\alpha)} \\ \downarrow{\scriptstyle f_{s(\alpha)}} & & \downarrow{\scriptstyle f_{t(\alpha)}} \\ P(i)_{s(\alpha)} & \xrightarrow{\varphi_\alpha} & P(i)_{t(\alpha)}. \end{array}$$

Let c be a nonconstant path from i to $s(\alpha)$. We have

$$f_{t(\alpha)}\varphi'_\alpha(c + I) = f_{t(\alpha)}(c\alpha + I) = c\alpha + I$$

is equal to

$$\varphi_\alpha f_{s(\alpha)}(c + I) = \varphi_\alpha(c + I) = c\alpha + I.$$

Since f is injective, it follows that M is a submodule of $P(i)$. Moreover, M is a maximal submodule since the quotient $P(i)/M \cong k$.

Now suppose that $L = (L_i, \psi_\alpha)$ is any proper submodule of $P(i)$, and let $h = (h_j)_{j \in Q_0} : L \to P(i)$ be an injective morphism. Then the basis element e_i in $P(i)_i$ does not lie in the image of h_i, because otherwise, for any path c starting at i, we would have

$$c = \varphi_c(e_i) = \varphi_c h_i(e_i) = h_{t(c)}\psi_c(e_i)$$

lies in the image of $h_{t(c)}$, and then L would not be a proper submodule of $P(i)$. Thus $\operatorname{im} h_i \subset P(i)'$, and we can define a morphism $g : L \to M$ by $g(x) = h(x)$ for all $x \in L$. This shows that L, and thus every proper submodule of $P(i)$, is a submodule of M, and thus $M = \operatorname{rad} P(i)$. □

Example 5.6. In Example 5.4, we have

$$\operatorname{rad} P(1) = {}^{2\,3}_{\;4}\ , \operatorname{rad} P(2) = 4\ , \operatorname{rad} P(3) = {}^{4}_{5}\ ,$$
$$P(4) = {}^{5}_{6}\ , \quad \operatorname{rad} P(5) = 6\ , \operatorname{rad} P(6) = 0\ .$$

Example 5.7. In Example 5.5, we have $\operatorname{rad} P(2) = 0$ and $\operatorname{rad} P(1)_1$ is the two-dimensional vector space with ordered basis $\{\alpha, \alpha^2\}$ and $\operatorname{rad} P(1)_2$ is the two-dimensional vector space with ordered basis $\{\beta, \alpha\beta\}$, and the linear maps are given with respect to these bases by the matrices:

$$\varphi_\alpha = \begin{bmatrix} 0 & 0 \\ 1 & 0 \end{bmatrix} \text{ and } \varphi_\beta = \begin{bmatrix} 0 & 0 \\ 1 & 0 \end{bmatrix}.$$

Example 5.8. If $A = kQ/I$ is a bound quiver algebra, then for every vertex i of Q, the projection $P(i) \to S(i)$ is a projective cover of the simple module $S(i)$, and we have a short exact sequence:

$$0 \longrightarrow \operatorname{rad} P(i) \longrightarrow P(i) \longrightarrow S(i) \longrightarrow 0.$$

5.4 Homological Dimensions

In this section, we define the projective dimension and the injective dimension of a module as well as the global dimension of an algebra. Roughly speaking, the projective dimension (respectively injective dimension) measures how far a module is from being projective (respectively injective), and the global dimension measures how far an algebra is from being hereditary.

Definition 5.5. Let M be an A-module. The **projective dimension** $\operatorname{pd} M$ of M is the smallest integer d such that there exists a projective resolution of the form

$$0 \longrightarrow P_d \longrightarrow P_{d-1} \longrightarrow \cdots \longrightarrow P_1 \longrightarrow P_0 \longrightarrow M \longrightarrow 0.$$

If no such resolution exists, then we say that M has infinite projective dimension.

Dually, the **injective dimension** $\operatorname{id} M$ of M is the smallest integer d such that there exists an injective resolution of the form

$$0 \longrightarrow M \longrightarrow I_0 \longrightarrow I_1 \longrightarrow \cdots \longrightarrow I_{d-1} \longrightarrow I_d \longrightarrow 0.$$

If no such resolution exists, then we say that M has infinite injective dimension.

The **global dimension** $\operatorname{gldim} A$ of the algebra A is defined as the supremum of the projective dimensions of all A-modules, that is,

$$\operatorname{gldim} A = \sup\{\operatorname{pd} M \mid M \in \operatorname{mod} A\}.$$

Remark 5.10. The global dimension of A can equivalently be defined as the supremum of the injective dimensions of all A-modules.

Remark 5.11. 1. A module M is projective if and only if $\operatorname{pd} M = 0$.
2. An algebra A is hereditary if and only if $\operatorname{gldim} A \leq 1$.

Example 5.9. The algebra given by the quiver

$$1 \xrightarrow{\alpha} 2 \xrightarrow{\beta} 3 \xrightarrow{\gamma} 4 \xrightarrow{\delta} 5$$

bound by $\alpha\beta = 0, \gamma\delta = 0$ is of global dimension 2. The fact that it is at least 2 can be seen by computing a minimal projective resolution for the simple module $S(1)$:

$$0 \longrightarrow P(3) \longrightarrow P(2) \longrightarrow P(1) \longrightarrow S(1) \longrightarrow 0,$$

which shows that $\operatorname{pd} S(1) = 2$.

Example 5.10. Now take the algebra given by the same quiver

$$1 \xrightarrow{\alpha} 2 \xrightarrow{\beta} 3 \xrightarrow{\gamma} 4 \xrightarrow{\delta} 5$$

but modify the relations to $\alpha\beta\gamma = 0, \gamma\delta = 0$. Then the minimal projective resolution for the simple module $S(1)$ becomes

$$0 \longrightarrow P(5) \longrightarrow P(4) \longrightarrow P(2) \longrightarrow P(1) \longrightarrow S(1) \longrightarrow 0,$$

which shows that $\operatorname{pd} S(1) = 3$. This algebra is of global dimension 3.

Example 5.11. The algebra given by the quiver

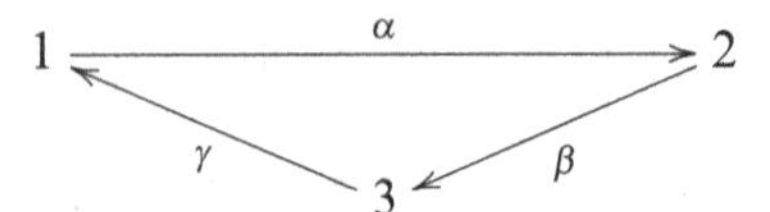

bound by $\alpha\,\beta = \beta\,\gamma = \gamma\,\alpha = 0$ is of infinite global dimension. Again, this can be seen by computing a minimal projective resolution for the simple module $S(1)$:

$$\cdots \longrightarrow P(3) \longrightarrow P(2) \longrightarrow P(1) \longrightarrow P(3) \longrightarrow P(2) \longrightarrow P(1) \longrightarrow S(1) \longrightarrow 0.$$

5.5 Auslander–Reiten Quivers of Bound Quiver Algebras

In this section, we give a few examples of Auslander–Reiten quivers for bound quiver algebras. The main difference to the hereditary path algebras is that an arrow in the Auslander–Reiten quiver that ends at a projective module does not necessarily start at a projective module; but it does always start at an indecomposable direct summand of the radical of the projective.

Example 5.12. Let Q be the quiver

$$1 \xrightarrow{\ \alpha\ } 2 \xrightarrow{\ \beta\ } 3 \xrightarrow{\ \gamma\ } 4$$

and $I = \langle \alpha\beta\gamma \rangle$. For this example, we use the knitting algorithm to construct the Auslander–Reiten quiver. The reader may follow the construction in the illustration in Fig. 5.1.

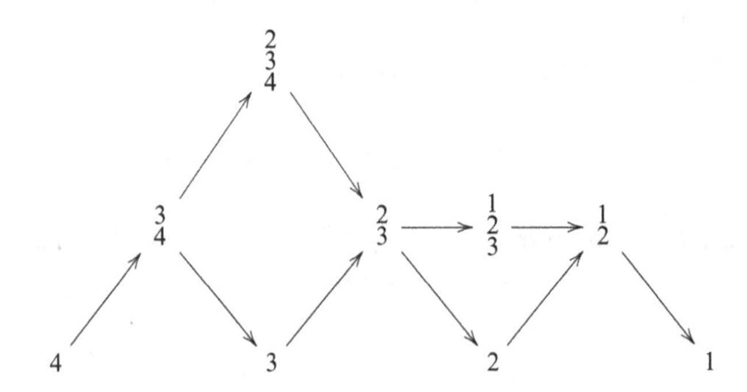

Fig. 5.1 The Auslander–Reiten quiver of Example 5.12

The projective modules are

$$P(1) = \begin{smallmatrix} 1 \\ 2 \\ 3 \end{smallmatrix}, \; P(2) = \begin{smallmatrix} 2 \\ 3 \\ 4 \end{smallmatrix}, \; P(3) = \begin{smallmatrix} 3 \\ 4 \end{smallmatrix}, \; P(4) = 4,$$

and the radicals of these projectives are

$$\operatorname{rad} P(1) = \begin{smallmatrix} 2 \\ 3 \end{smallmatrix}, \; \operatorname{rad} P(2) = \begin{smallmatrix} 3 \\ 4 \end{smallmatrix}, \; \operatorname{rad} P(3) = 4, \; \operatorname{rad} P(4) = 0.$$

Start with a simple projective $P(4)$, draw an arrow to the projective $P(3)$, since $P(4)$ is a direct summand of rad $P(3)$, and then draw an arrow from $P(3)$ to $P(2)$, since $P(3)$ is a direct summand of rad $P(2)$.

Now we have started three τ-orbits and we use the knitting algorithm until we get the module $\begin{smallmatrix} 2 \\ 3 \end{smallmatrix}$ which is the radical of the remaining projective $P(1)$.

At this point we must start a new τ-orbit for $P(1)$ with an arrow rad $P(1) \to P(1)$. Since rad $P(1)$ appears on the middle τ-orbit, we insert this new τ-orbit to the right of rad $P(i)$ as in the Auslander–Reiten quivers of type $\mathbb{D}_n$.

Then continue the knitting algorithm until all injectives are obtained.

Example 5.13. Let Q be the quiver

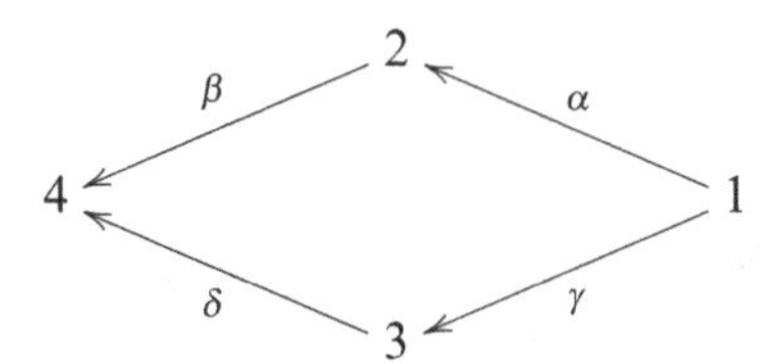

and $I = \langle \alpha\beta - \gamma\delta \rangle$. In this example, we can also use the knitting algorithm to construct the Auslander–Reiten quiver. The projective modules are

$$\begin{smallmatrix} & 1 & \\ 2 & & 3 \\ & 4 & \end{smallmatrix}, \quad \begin{smallmatrix} 2 \\ 4 \end{smallmatrix}, \quad \begin{smallmatrix} 3 \\ 4 \end{smallmatrix}, \quad 4.$$

Note that rad $P(2) =$ rad $P(3) = P(4)$ and that rad $P(1)$ is not projective.

The Auslander–Reiten quiver of kQ/I is given by

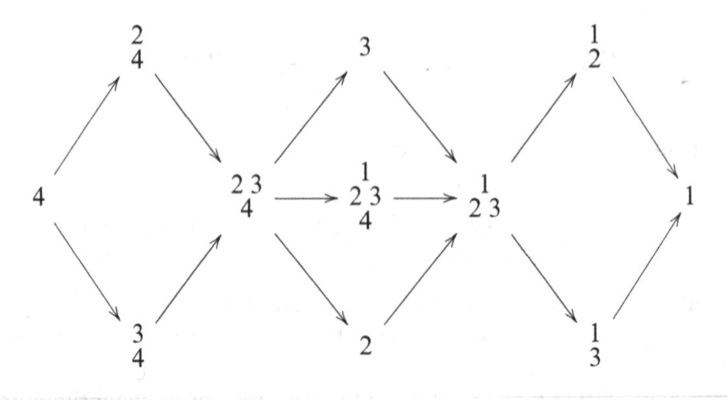

Example 5.14. Let Q be the quiver

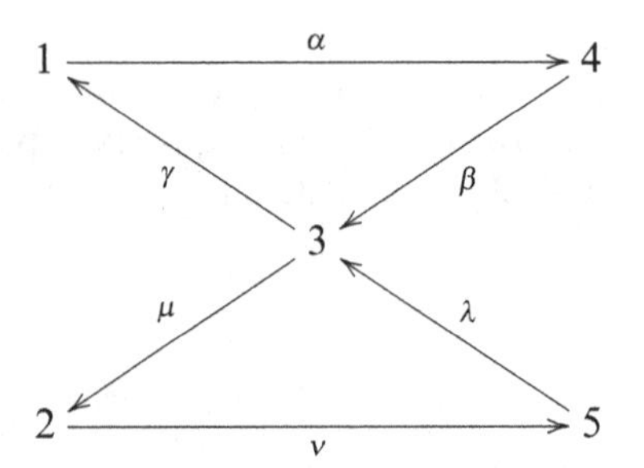

bound by the relations $\alpha\,\beta = 0$, $\beta\,\gamma = 0$, $\gamma\,\alpha = 0$ $\lambda\,\mu = 0$, $\mu\,\nu = 0$, and $\nu\,\lambda = 0$.

The quiver in this example has oriented cycles and the Auslander–Reiten quiver of the bound quiver algebra also contains oriented cycles. We first compute the projectives and their radicals:

$$P(1) = \begin{smallmatrix}1\\4\end{smallmatrix}\,,\ P(2) = \begin{smallmatrix}2\\5\end{smallmatrix}\,,\ P(3) = \begin{smallmatrix}&3&\\1&&2\end{smallmatrix}\,,\ P(4) = \begin{smallmatrix}4\\3\\2\end{smallmatrix}\,,\ P(5) = \begin{smallmatrix}5\\3\\1\end{smallmatrix}$$

$$\operatorname{rad} P(1) = 4\ ,\ \operatorname{rad} P(2) = 5\ ,\ \operatorname{rad} P(3) = 1 \oplus 2,$$

$$\operatorname{rad} P(4) = \begin{smallmatrix}3\\2\end{smallmatrix}\,,\ \operatorname{rad} P(5) = \begin{smallmatrix}3\\1\end{smallmatrix}$$

Now we start the knitting algorithm with the projective $P(3)$ and its two simple radical summands:

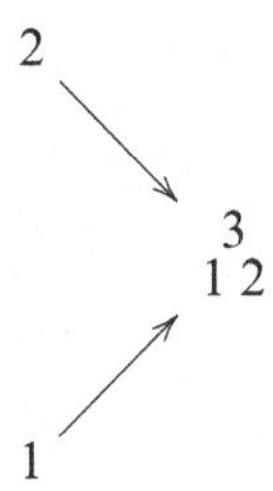

The algorithm gives then the radicals of $P(4)$ and $P(5)$, so we introduce two new τ-orbits for these new projectives, and continue knitting:

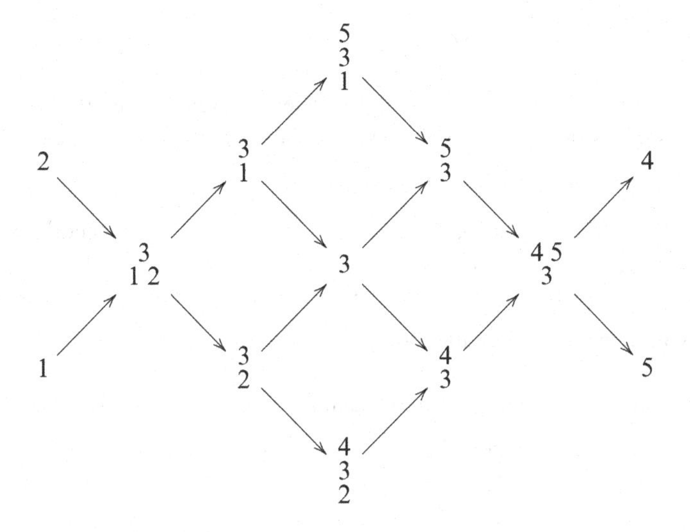

Now we have the radicals of $P(1)$ and $P(2)$, so we introduce new τ-orbits for them and continue knitting. We then get back to the modules we have started with, and the Auslander–Reiten quiver is complete. One has to identify the vertices that have the same label, thus creating a Moebius strip:

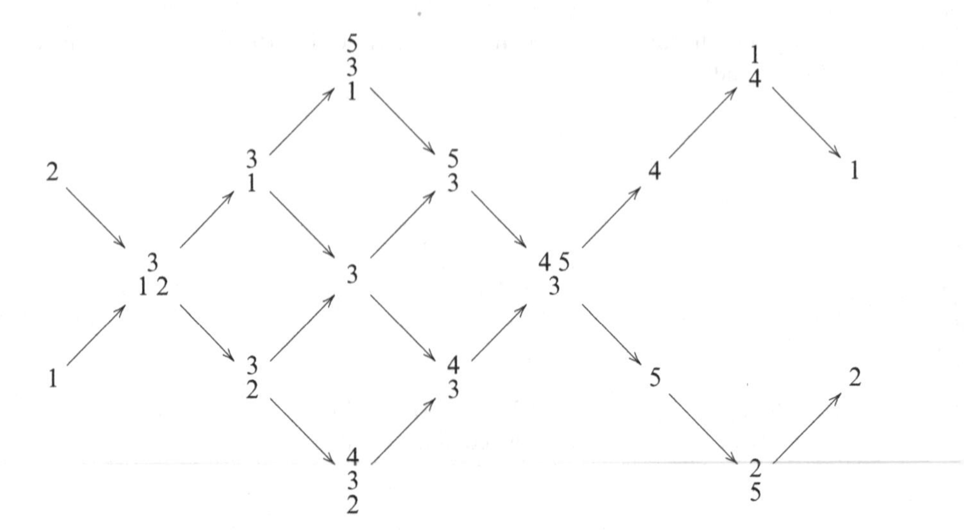

Remark 5.12. Let us point out that if Q has oriented cycles, then the above method will not always work. Even in the example above, if we had started with the projective $P(5)$ and its radical $\begin{smallmatrix}3\\1\end{smallmatrix}$, then the knitting algorithm would predict the simple $S(5)$ as $\tau^{-1}\mathrm{rad}\,P(5)$, but this is wrong! One should always check that the first mesh is correctly computed. This can be done by calculating the Auslander–Reiten translation τ^{-1} of the radical of the projective. In our example, this can be done using the following diagram, where the top row is a minimal injective presentation of $\begin{smallmatrix}3\\1\end{smallmatrix}$, and the bottom row is the projective presentation of $\tau^{-1}\begin{smallmatrix}3\\1\end{smallmatrix}$ obtained by applying the Nakayama functor as in Sect. 2.3:

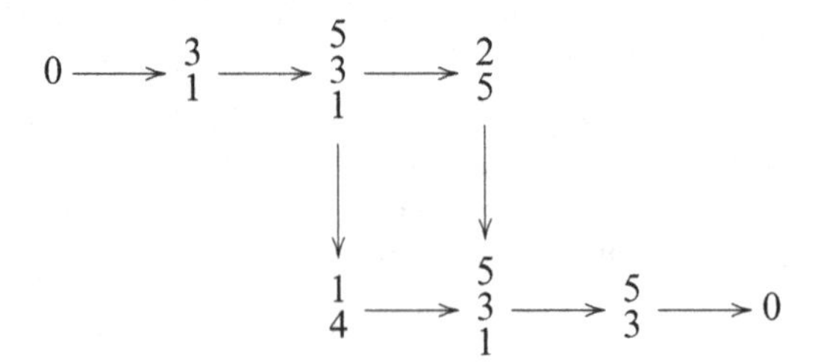

Thus $\tau^{-1}\begin{smallmatrix}3\\1\end{smallmatrix} = \begin{smallmatrix}5\\3\end{smallmatrix}$.

Example 5.15. To illustrate the fact that the Auslander–Reiten quivers are not always that easy to obtain, even in finite representation type, we give an example where the number of τ-orbits is greater than the number of indecomposable projectives.

Let Q be the quiver

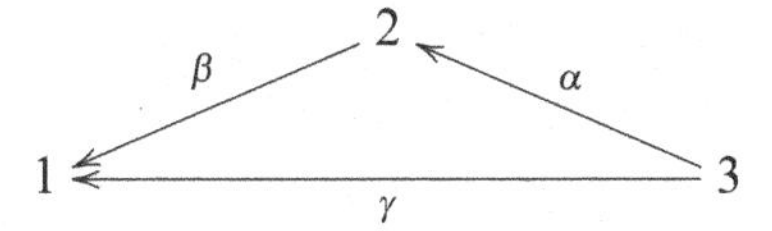

bound by the relation $\alpha\beta = 0$. Then

$$P(1) = 1\,,\ P(2) = \begin{smallmatrix} 2 \\ 1 \end{smallmatrix}\,,\ P(3) = \begin{smallmatrix} & 3 & \\ 2 & & 1 \end{smallmatrix}$$

$$\operatorname{rad} P(1) = 0\,,\ \operatorname{rad} P(2) = 1\,,\ \operatorname{rad} P(3) = 2 \oplus 1$$

Here $P(1)$ is a direct summand of the radicals of $P(2)$ and $P(3)$; thus we have arrows $P(1) \to P(2)$ and $P(1) \to P(3)$ in the Auslander–Reiten quiver. Moreover, the simple module $S(2)$ is a direct summand of the radical of $P(3)$, so we get an arrow $S(2) \to P(3)$ which gives rise to a fourth τ-orbit. Using the knitting algorithm from here we get the whole Auslander–Reiten quiver below, where we have to identify the two vertices labeled 2:

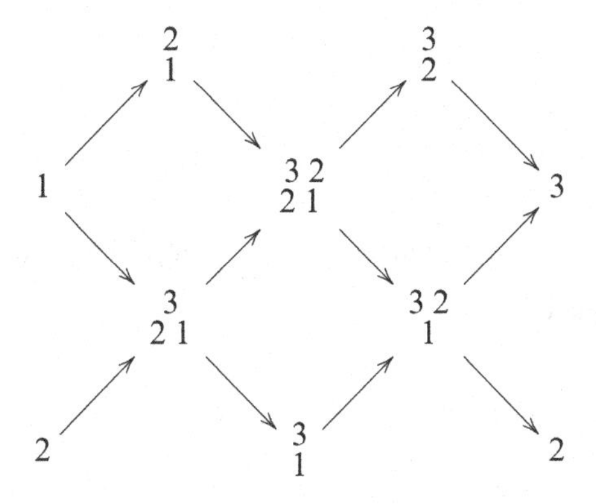

Note for example that $\operatorname{Hom}(\begin{smallmatrix} 2 \\ 1 \end{smallmatrix}, \begin{smallmatrix} 3 & 2 \\ 2 & 1 \end{smallmatrix})$ is of dimension 2, since there is the irreducible injective morphism that corresponds to the arrow in the Auslander–Reiten quiver and there is also the morphism with simple kernel $S(1)$ and simple image $S(2)$ corresponding to the lower left 2 in $\begin{smallmatrix} 3 & 2 \\ 2 & 1 \end{smallmatrix}$. This morphism is given in the Auslander–Reiten quiver as the composition of 5 arrows, starting at $\begin{smallmatrix} 2 \\ 1 \end{smallmatrix}$ going downward until $S(2)$ and then, using the identification in the Auslander-Reiten quiver, going upward until $\begin{smallmatrix} 3 & 2 \\ 2 & 1 \end{smallmatrix}$.

Problems

Exercises for Chapter 5

5.1. Let Q be the quiver

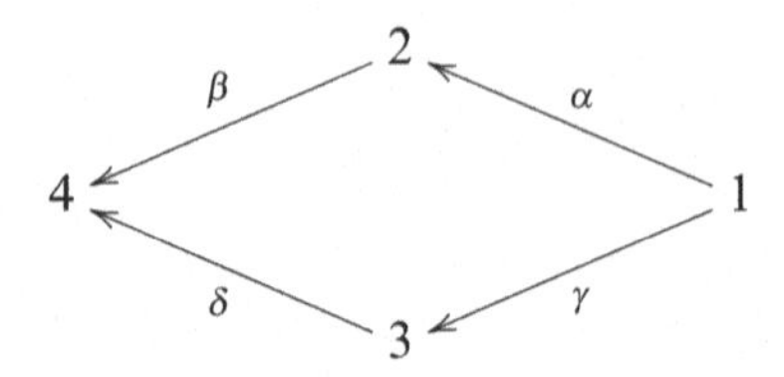

and define two ideals $I_1 = \langle\alpha\beta + \gamma\delta\rangle$ and $I_2 = \langle\alpha\beta - \gamma\delta\rangle$. Show that

1. $I_1 \neq I_2$ unless the characteristic of k is 2 and
2. there exists an isomorphism of algebras

$$kQ/I_1 \longrightarrow kQ/I_2.$$

5.2. Let Q be the quiver

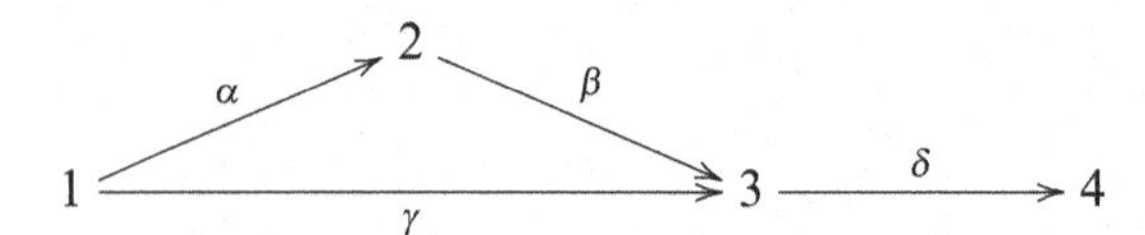

and define two ideals $I_1 = \langle\gamma\delta\rangle$ and $I_2 = \langle\gamma\delta - \alpha\beta\delta\rangle$. Show that $kQ/I_1 \cong kQ/I_2$.

5.3. Let Q be the quiver

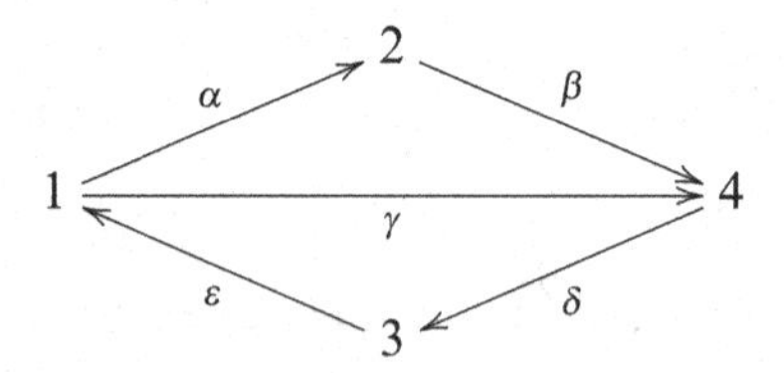

and define two ideals $I_1 = \langle\gamma\delta, \delta\epsilon\rangle$ and $I_2 = \langle\gamma\delta - \alpha\beta\delta, \delta\epsilon\rangle$. Show that $kQ/I_1 \cong kQ/I_2$. Show that $kQ/I_1 \cong kQ/I_2$.

5.4. Prove that the modules $P(i)$ in Definition 5.3 are projective and indecomposable.

5.5. Prove that Theorem 2.11 holds for bound quiver algebras.

5.6. The goal of this exercise is to generalize the discussion of the Nakayama functor in Sect. 2.3 to bound quiver algebras.

Let $A = kQ/I$ be a bound quiver algebra, $D = \mathrm{Hom}_A(-, k)$ the duality, and $i \in Q_0$ be a vertex:

1. Show that $D: \mathrm{mod}\, A \to \mathrm{mod}\, A^{\mathrm{op}}$ is a duality of categories which induces a duality $D: \mathrm{proj}\, A \to \mathrm{inj}\, A^{\mathrm{op}}$ which maps $P_A(i)$ to $I_{A^{\mathrm{op}}}(i)$.
2. Show that $\mathrm{Hom}_A(-, A): \mathrm{proj}\, A \to \mathrm{proj}\, A^{\mathrm{op}}$ is a duality which maps $P_A(i)$ to $P_{A^{\mathrm{op}}}(i)$.
3. Conclude that the Nakayama functor ν induces an equivalence of categories $\mathrm{proj}\, A \to \mathrm{inj}\, A$ which maps $P_A(i)$ to $I_A(i)$.

5.7. Give an example of a bound quiver algebra of global dimension 3.

5.8. Give an example of a bound quiver algebra of global dimension 4.

5.9. Give an example of a bound quiver algebra of infinite global dimension.

5.10. An algebra is called *1-Gorenstein* if each projective module has injective dimension 1, and each injective module has projective dimension 1. Prove that the algebra in Example 5.14 is 1-Gorenstein.

Chapter 6
New Algebras from Old

In this chapter, we present several popular constructions for algebras, each one in a separate section. We introduce tilted algebras, trivial extensions, cluster-tilted algebras, triangular matrix algebras, and one-point extensions.

A *tilted algebra* is the endomorphism algebra $\mathrm{End}_A T$ of a special type of module T, called tilting module, over a hereditary algebra A. By definition, a tilting module T satisfies two conditions. First, T has no self-extensions, that is, $\mathrm{Ext}^1(T, T) = 0$. Second, T is maximal in the sense that any indecomposable module which is not a direct summand of T has an extension with T. The type of a tilted algebra is the type of the underlying hereditary algebra A. We construct several examples of tilted algebras of types $\mathbb{A}$ and $\mathbb{D}$.

The *trivial extension* algebras are constructed from an algebra A and an A-bimodule M by introducing a multiplication on the direct sum $A \oplus M$. In the special case where $M = DA$ is the dual of the algebra, the trivial extension is a symmetric algebra, which means that the algebra is isomorphic to its dual as a bimodule. In particular, every projective module is also injective.

Cluster-tilted algebras are trivial extensions of a tilted algebra C with respect to the bimodule $M = \mathrm{Ext}^2_C(DC, C)$. If the tilted algebra is of type $\mathbb{A}$ or $\mathbb{D}$, then the corresponding cluster-tilted algebra is given by the bound quiver of a triangulation of a polygon or punctured polygon, respectively. We have seen several examples of these cluster-tilted algebras in Chap. 3.

A *triangular matrix* algebra is a special kind of trivial extension algebra having the property that its elements can be written as lower triangular 2 by 2 matrices in such a way that the product in the algebra is given by matrix multiplication. The elements which are zero except at the diagonal position (1,1) form a quotient algebra A_1, and the elements which are zero except at the diagonal position (2,2) form a quotient algebra A_2.

Finally, a *one-point extension* is a special kind of triangular matrix algebra for which the algebra A_2 is equal to the field k. The name one-point extension stems from the fact that quiver of a one-point extension has one vertex more than the quiver of its quotient algebra A_1.

R. Schiffler, *Quiver Representations*, CMS Books in Mathematics,
DOI 10.1007/978-3-319-09204-1_6

6.1 Tilted Algebras

Let Q be a quiver without oriented cycles, let n be the number of vertices, and let $A = kQ$ be the path algebra.

Definition 6.1. An A-module T is called a **tilting module** if

1. $T = T_1 \oplus \cdots \oplus T_n$, where the T_i are pairwise non-isomorphic indecomposable A-modules, and
2. $\mathrm{Ext}^1(T, T) = 0$.

Example 6.1. The results in Chap. 3 allow us to find all the tilting modules for path algebras of type $\mathbb{A}$ and $\mathbb{D}$.

For example, if Q is the quiver $1 \longrightarrow 2 \longrightarrow 3$, then the Auslander-Reiten quiver is

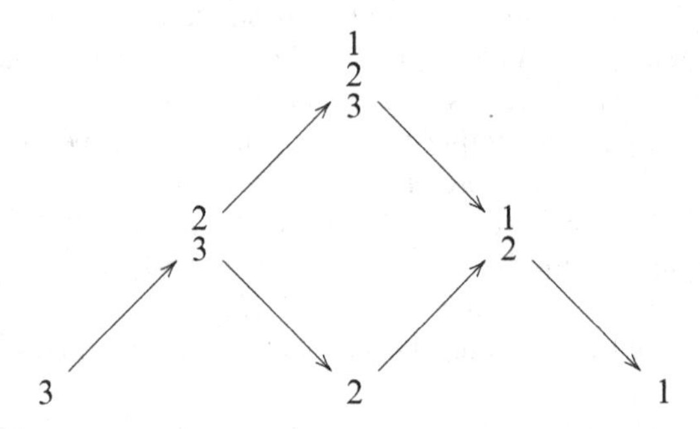

Now suppose that the simple module $S(3) = 3$ is an indecomposable summand of a tilting module T. Since $S(3)$ is projective, $\mathrm{Ext}^1(S(3), M)$ is zero for all modules M, and, on the other hand, we know from Chap. 3 that $\mathrm{Ext}^1(M, S(3)) \neq 0$ if and only if $M \in \left\{2, \begin{smallmatrix}1\\2\end{smallmatrix}\right\}$. Thus the other two summands of T must be among

$$\begin{smallmatrix}2\\3\end{smallmatrix}, \begin{smallmatrix}1\\2\\3\end{smallmatrix} \text{ and } 1\,.$$

Since $\mathrm{Ext}^1(1, \begin{smallmatrix}2\\3\end{smallmatrix}) \neq 0$, we see that the only tilting modules containing $S(3)$ are

$$3 \oplus \begin{smallmatrix}2\\3\end{smallmatrix} \oplus \begin{smallmatrix}1\\2\\3\end{smallmatrix} \quad \text{and} \quad 3 \oplus \begin{smallmatrix}1\\2\\3\end{smallmatrix} \oplus 1\,.$$

Next consider the tilting modules that have $S(2)$ as a direct summand. This excludes $S(1)$ and $S(3)$ and we get the two tilting modules:

$$\begin{smallmatrix}2\\3\end{smallmatrix} \oplus 2 \oplus \begin{smallmatrix}1\\2\\3\end{smallmatrix} \quad \text{and} \quad 2 \oplus \begin{smallmatrix}1\\2\\3\end{smallmatrix} \oplus \begin{smallmatrix}1\\2\end{smallmatrix}.$$

Finally, there is one tilting module containing neither $S(1)$ nor $S(2)$ given by

$$\begin{smallmatrix}1\\2\\3\end{smallmatrix} \oplus \begin{smallmatrix}1\\2\end{smallmatrix} \oplus 1\,.$$

Note that in this example, the tilting module

$$3 \oplus \begin{smallmatrix}1\\2\\3\end{smallmatrix} \oplus 1$$

is the only one that does not form a connected subquiver in the Auslander–Reiten quiver.

Example 6.2. Let Q be the quiver

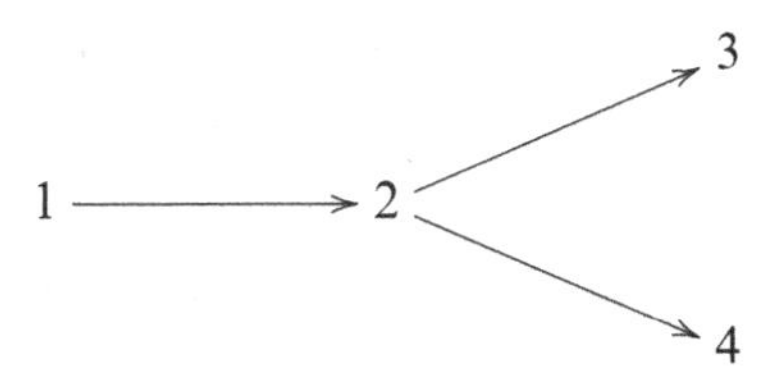

and let $A = kQ$ be its path algebra. Then the Auslander–Reiten quiver is

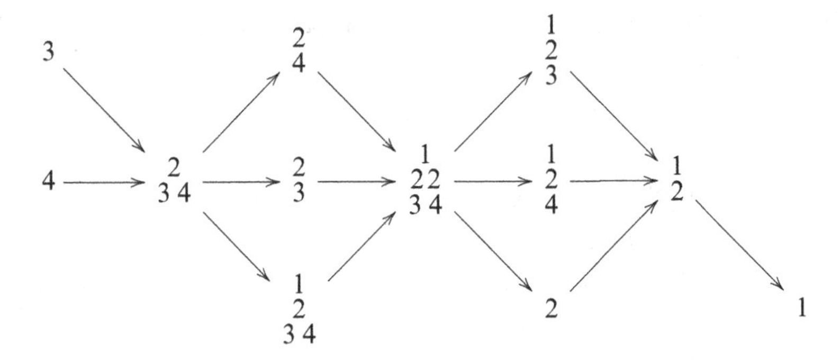

Consider tilting modules that have $S(3) = 3$ as a direct summand. We have seen in Chap. 3 that $\mathrm{Ext}^1(M, S(3)) \neq 0$ if M is in the hammock starting at $\tau^{-1}S(3)$. Thus the other direct summands of the tilting module must be taken from

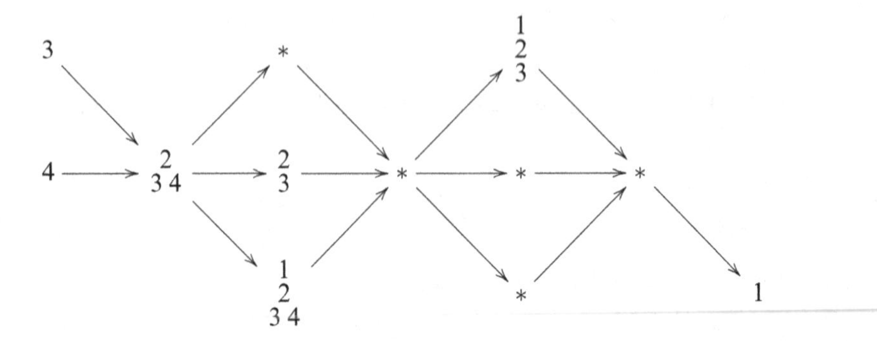

If $P(2)$ is also a direct summand, then there are the two tilting modules:

$$3 \oplus \begin{smallmatrix} 2 \\ 3\,4 \end{smallmatrix} \oplus 4 \oplus \begin{smallmatrix} 1 \\ 2 \\ 3\,4 \end{smallmatrix} \quad \text{and} \quad 3 \oplus \begin{smallmatrix} 2 \\ 3\,4 \end{smallmatrix} \oplus \begin{smallmatrix} 2 \\ 3 \end{smallmatrix} \oplus \begin{smallmatrix} 1 \\ 2 \\ 3\,4 \end{smallmatrix}.$$

These tilting modules induce connected subquivers in the Auslander–Reiten quiver.

Now suppose that $P(2)$ is not a direct summand of the tilting module. Then the only possibilities are

$$3 \oplus 4 \oplus \begin{smallmatrix} 1 \\ 2 \\ 3\,4 \end{smallmatrix} \oplus 1\,, \; 3 \oplus \begin{smallmatrix} 2 \\ 3 \end{smallmatrix} \oplus \begin{smallmatrix} 1 \\ 2 \\ 3\,4 \end{smallmatrix} \oplus \begin{smallmatrix} 1 \\ 2 \\ 3 \end{smallmatrix} \quad \text{and} \quad 3 \oplus \begin{smallmatrix} 1 \\ 2 \\ 3\,4 \end{smallmatrix} \oplus \begin{smallmatrix} 1 \\ 2 \\ 3 \end{smallmatrix} \oplus 1.$$

These tilting modules do not induce connected subgraphs in the Auslander–Reiten quiver.

Definition 6.2. Let T be a tilting module over the path algebra $A = kQ$. Then the endomorphism algebra

$$\mathrm{End}_A T$$

is called a **tilted algebra** of type Q.

So a tilted algebra B is of the form

$$B = \mathrm{End}_A T = \bigoplus_{i,j=1}^{n} \mathrm{Hom}(T_i, T_j),$$

where T_i, T_j are indecomposable summands of T. The unity element of B is given by the identity morphism $1_T: T \to T$. Clearly, this morphism can be decomposed with respect to the direct sum as $1_T = 1_{T_1} + \cdots + 1_{T_n}$, where 1_{T_i} is the identity morphism on the module T_i (extended by zero to the T_j with $j \neq i$). Moreover, the 1_{T_i} are primitive idempotents which are pairwise orthogonal. Thus the quiver of B has n vertices corresponding to the n indecomposable summands $T_1, \ldots, T_n$. There is an arrow $i \to j$ in this quiver if and only if

- $i \neq j$,
- there is a nonzero morphism of A-modules $f: T_j \to T_i$, and
- f does not factor nontrivially through one of the T_ℓ.

If the dimension of $\mathrm{Hom}(T_j, T_i)$ is greater than one, then we may have more than one arrow from i to j in the quiver of B; the number of arrows is the dimension of the quotient space of $\mathrm{Hom}(T_j, T_i)$ by the subspace of all morphisms that factor nontrivially through one of the T_ℓ. Paths in the quiver of B correspond to compositions of such morphisms, and the relations for the quiver of B reflect the relations among these morphisms.

Example 6.3. We have seen that the right A-module A decomposes into the direct sum of the indecomposable projective modules $A = \oplus_{i=1}^n P(i)$. Since $\mathrm{Ext}^1(P, -)$ is zero for every projective module P, then, in particular, $\mathrm{Ext}^1(A, A) = 0$, and we see that A is a tilting module. The corresponding tilted algebra is isomorphic to A, that is,

$$A \cong \mathrm{End}_A A.$$

This shows that *hereditary algebras are tilted algebras.*

Example 6.4. Let $T = T_1 \oplus T_2 \oplus T_3$ be the tilting module of Example 6.1 given by

$$T_1 = 3 \quad , \quad T_2 = \begin{smallmatrix} 1 \\ 2 \\ 3 \end{smallmatrix} \quad , \quad T_3 = 1 \, ,$$

and let $B = \mathrm{End}\, T$ be the tilted algebra. Then

$$B = \begin{bmatrix} \mathrm{Hom}(T_1, T_1) & \mathrm{Hom}(T_2, T_1) & \mathrm{Hom}(T_3, T_1) \\ \mathrm{Hom}(T_1, T_2) & \mathrm{Hom}(T_2, T_2) & \mathrm{Hom}(T_3, T_2) \\ \mathrm{Hom}(T_1, T_3) & \mathrm{Hom}(T_2, T_3) & \mathrm{Hom}(T_3, T_3) \end{bmatrix} = \begin{bmatrix} k & 0 & 0 \\ k & k & 0 \\ 0 & k & k \end{bmatrix},$$

and therefore, B is the bound quiver algebra given by the quiver

$$1 \xleftarrow{\ \beta\ } 2 \xleftarrow{\ \alpha\ } 3$$

and relation $\alpha\beta$.

The other tilting modules of Example 6.1 give rise to tilted algebras that are path algebras of type $\mathbb{A}_3$ without relations.

Example 6.5. Now consider the tilting modules of Example 6.2. The tilting modules that induce connected subquivers in the Auslander–Reiten quiver of A again give rise to tilted algebras that are path algebras of type $\mathbb{D}_4$ without relations. Let us consider the other three examples.

If $T = T_1 \oplus \cdots \oplus T_4$ with

$$T_1 = 3 \quad , \quad T_2 = 4 \quad , \quad T_3 = \begin{smallmatrix} 1 \\ 2 \\ 3\,4 \end{smallmatrix} \quad , \quad T_4 = 1,$$

then $\operatorname{End} T$ is given by the quiver

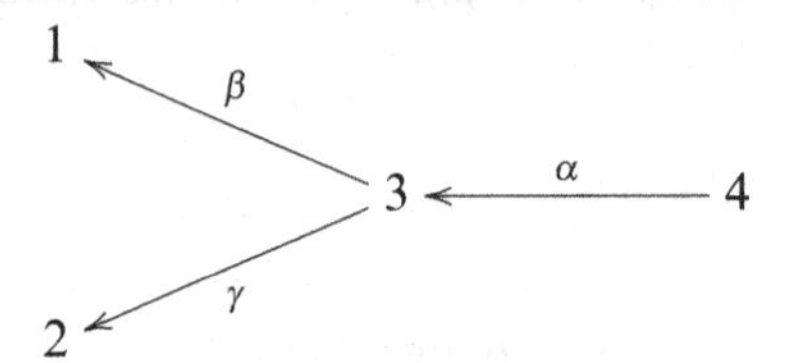

and relations $\alpha\beta = 0$, $\alpha\gamma = 0$, because the composition of the morphisms $T_1 \to T_3 \to T_4$ is zero and the composition $T_2 \to T_3 \to T_4$ is zero as well.

If $T = T_1 \oplus \cdots \oplus T_4$ with

$$T_1 = 3 \quad , \quad T_2 = \begin{smallmatrix} 2 \\ 3 \end{smallmatrix} \quad , \quad T_3 = \begin{smallmatrix} 1 \\ 2 \\ 3\,4 \end{smallmatrix} \quad , \quad T_4 = \begin{smallmatrix} 1 \\ 2 \\ 3 \end{smallmatrix},$$

then $\operatorname{End} T$ is given by the quiver

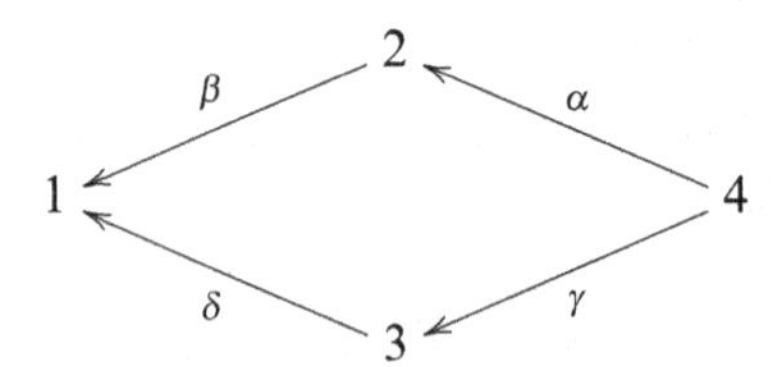

and relations $\alpha\beta - \gamma\delta = 0$, because the composition of the morphisms $T_1 \to T_2 \to T_4$ is equal to the composition $T_1 \to T_3 \to T_4$.

Finally, if $T = T_1 \oplus \cdots \oplus T_4$ with

$$T_1 = 3 \quad , \quad T_2 = \begin{smallmatrix} 1 \\ 2 \\ 3\,4 \end{smallmatrix} \quad , \quad T_3 = \begin{smallmatrix} 1 \\ 2 \\ 3 \end{smallmatrix} \quad , \quad T_4 = 1,$$

then End T is given by the quiver

$$1 \xleftarrow{\gamma} 2 \xleftarrow{\beta} 3 \xleftarrow{\alpha} 4$$

and relation $\alpha\beta\gamma = 0$, because the compositions of the morphisms $T_1 \to T_2 \to T_3$ and $T_2 \to T_3 \to T_4$ are nonzero, but the composition $T_1 \to T_2 \to T_3 \to T_4$ is zero.

6.2 Trivial Extensions

Let A be a finite-dimensional k-algebra and let M be an A-bimodule, that is, M is a right A-module and a left A-module and $(am)a' = a(ma')$ for all $a, a' \in A, m \in M$. The **trivial extension** $A \ltimes M$ is the k-algebra whose vector space is the direct sum $A \oplus M$ and whose multiplication is given by the formula

$$(a, m)(a', m') = (aa', am' + ma').$$

Proposition 6.1. *Let A and M be as above. Then*

1. $1_{A\ltimes M} = (1_A, 0)$.
2. *The map $i : A \to A \ltimes M$ given by $i(a) = (a, 0)$ is an injective algebra homomorphism.*
3. *The map $\pi : A \ltimes M \to A$ given by $\pi(a, m) = a$ is a surjective algebra homomorphism, and $\pi \circ i = 1_A$.*
4. *The bimodule $M \cong \{(0, m) \mid m \in M\}$ is a two-sided nilpotent ideal in $A \ltimes M$ and $M^2 = 0$.*
5. $\operatorname{rad}(A \ltimes M) = \operatorname{rad} A + M$.
6. *There is a short exact sequence of $(A \ltimes M)$-bimodules:*

$$0 \longrightarrow M \longrightarrow A \ltimes M \longrightarrow A \longrightarrow 0\,.$$

For every vertex i, this sequence induces a short exact sequence of right $A \ltimes M$-modules:

$$0 \longrightarrow e_iM \longrightarrow P(i) \longrightarrow P_A(i) \longrightarrow 0\,,$$

where $P(i)$ denotes the indecomposable projective $(A \ltimes M)$-module and $P_A(i)$ is the indecomposable projective A-module considered as $(A \ltimes M)$-module.

Proof. (1)–(3) are clear from the definition.

(4) The fact that $M = \ker \pi$ shows that M is a two-sided ideal. Moreover $(0, m)(0, m') = (0, 0)$, which shows that $M^2 = 0$, and M is nilpotent.

(5) From Lemma 4.1, we know that $(a, m) \in \operatorname{rad}(A \ltimes M)$ if and only if for all $(a', m') \in A \ltimes M$, the element

$$(1, 0) - (a, m)(a', m') = (1 - aa', -(am' + ma'))$$

has a right inverse. This holds if and only if for all (a', m'), there exists (c, n) such that

$$(1, 0) = ((1 - aa')c, (1 - aa')n - (am' + ma')c). \tag{6.1}$$

In particular, for all $a' \in A$, the element $(1 - aa')$ has a right inverse c in the algebra A, which is equivalent to $a \in \operatorname{rad} A$, again by Lemma 4.1. This shows that if $(a, m) \in \operatorname{rad}(A \ltimes M)$, then $a \in \operatorname{rad} A$; thus $\operatorname{rad}(A \ltimes M) \subseteq \operatorname{rad} A + M$.

Moreover, if $1 = (1 - aa')c$, then the element $n = c(am' + ma')c \in M$ is a solution for equation (6.1), and therefore $\operatorname{rad}(A \ltimes M) \supseteq \operatorname{rad} A + M$.

(6) It follows from (4) that M is an $(A \ltimes M)$-bimodule, so the inclusion $M \to A \ltimes M$ is a morphism which gives rise to the first short exact sequence. The second short exact sequence is obtained by multiplying with the idempotent e_i. □

We will consider several special cases of this construction in the following subsections.

6.3 Self-Injective Algebras and the Trivial Extensions $A \ltimes DA$

In this subsection, we consider the trivial extension of A by the A-bimodule DA.

Definition 6.3. A k-algebra B is called **selfinjective** if the right B-module is projective and injective.

The trivial extension algebra $A \ltimes DA$ is a selfinjective algebra. Before proving this result, let us develop some results about selfinjective algebras.

Proposition 6.2. *The following are equivalent:*

1. *B is selfinjective.*
2. *A B-module is projective if and only if it is injective.*

Proof. (1) ⇒ (2) Let M be an indecomposable B-module. If M is projective, then M is a direct summand of B, but then M is also injective because B is self-injective. Since the number of indecomposable projective modules is equal to the number of indecomposable injective modules, this shows that an indecomposable B-module is projective if and only if it is injective. The result now follows from Proposition 2.7.

(2) ⇒ (1) The B-module B is projective, so it is injective by (2), and the algebra B is selfinjective. □

Next we need the concept of syzygies. Let B be an algebra and fix a projective cover $P(M) \to M$ for each B-module M. Then the **syzygy** ΩM of M is the kernel of that projective cover. We thus have a short exact sequence:

$$0 \longrightarrow \Omega M \longrightarrow P(M) \longrightarrow M \longrightarrow 0.$$

Note that the syzygy of a module is zero if and only if the module is projective.

If M is not projective, then ΩM is again a nonzero B-module, and we can compute its syzygy $\Omega^2 M$. Iterating this procedure, the ith syzygy $\Omega^i M$ is the kernel of the morphism f_{i-1} in a minimal projective resolution:

$$\cdots \longrightarrow P_j \xrightarrow{f_j} \cdots \xrightarrow{f_2} P_1 \xrightarrow{f_1} P_0 \xrightarrow{f_0} M \longrightarrow 0\,.$$

The next proposition describes the Auslander–Reiten translation in a self-injective algebra as the composition of the Nakayama functor ν and Ω^2.

Proposition 6.3. *Let B be a selfinjective k-algebra, and let M be a B-module. Then*

$$\tau M \cong \Omega^2 \nu M.$$

Proof. To compute τM, we take a minimal projective resolution

$$\cdots \longrightarrow P_1 \longrightarrow P_0 \longrightarrow M \longrightarrow 0$$

and apply the Nakayama functor to get an exact sequence:

$$0 \longrightarrow \tau M \longrightarrow \nu P_1 \longrightarrow \nu P_0 \longrightarrow \nu M \longrightarrow 0\,.$$

The Nakayama functor induces an equivalence proj $B \to$ inj B; see Proposition 2.29 and Exercise 5.6, and since B is selfinjective, the injective modules νP_i are also projective, and

$$\cdots \longrightarrow \nu P_1 \longrightarrow \nu P_0 \longrightarrow \nu M \longrightarrow 0$$

is a minimal projective resolution of νM. Thus $\tau M \cong \Omega^2 \nu M$. □

In this proof, we have seen that, for a self-injective algebra, the Nakayama functor ν maps indecomposable projective modules to indecomposable projective modules. Thus its restriction to the subcategory proj B of mod B whose objects are the projective B-modules gives an equivalence:

$$\nu|_{\text{proj}\, B} \colon \text{proj}\, B \longrightarrow \text{proj}\, B.$$

In the nicest case, when this equivalence is the identity, the algebra B is called symmetric; the formal definition is the following:

Definition 6.4. An algebra B is called **symmetric** if there is an isomorphism of B-bimodules $B \cong DB$.

Proposition 6.4. *Let B be a symmetric algebra. Then*

1. *B is selfinjective,*
2. $\nu \cong 1$,
3. $\tau = \Omega^2$.

Proof. Let $e_1, \ldots, e_n \in B$ be primitive, orthogonal idempotents such that $1_B = e_1 + \cdots + e_n$. Then by the results of Sect. 4.4, the right B-module B decomposes into indecomposable projective right B-modules as follows:

$$B = e_1 B \oplus \cdots \oplus e_n B.$$

The Nakayama functor maps the indecomposable projective $e_i B$ to the indecomposable injective $D(Be_i) = e_i DB$. Since B and DB are isomorphic as bimodules, multiplying on the left by e_i yields isomorphic right B-modules $e_i B \cong e_i DB$, and the right-hand side is equal to $\nu(e_i B)$. This shows that the Nakayama functor is isomorphic to the identity on proj B.

Now let M be a B-module, and let

$$\cdots \longrightarrow P_1 \xrightarrow{f} P_0 \longrightarrow M \longrightarrow 0$$

be a minimal projective resolution. Applying the right exact functor ν to this sequence, we get an exact sequence:

$$\cdots \longrightarrow \nu P_1 \xrightarrow{\nu f} \nu P_0 \longrightarrow \nu M \longrightarrow 0.$$

Thus $\nu M = \operatorname{coker} \nu f$, and, using the fact that ν is isomorphic to the identity on proj B, we get $\nu M \cong M$, and we have shown (2).

On the other hand, by Proposition 2.29, the Nakayama functor induces an equivalence proj $B \to$ inj B, and hence we have also shown that a module is projective if and only if it is injective, and then Proposition 6.2 implies (1).

Condition (3) follows from Proposition 6.3. □

Now let us return to our trivial extensions.

Proposition 6.5. *The trivial extension algebra $A \ltimes DA$ is symmetric.*

Proof. We will think of the elements of the dual as morphisms to the field k. Recall that the bimodule structure on DA is given as follows: Let $f : A \to k$ be an element of DA and let $a \in A$. Then af and fa are the elements of DA defined by

$$(af)(b) = f(ba) \quad \text{and} \quad (fa)(b) = f(ab).$$

We define a map

$$\phi: A \ltimes DA \to D(A \ltimes DA)$$

as follows: for every $(a, f) \in A \ltimes DA$, we define a morphism $\phi(a, f): A \ltimes DA \to k$ by $(\phi(a, f))(b, g) = f(b) + g(a)$.

We will show first that ϕ is a morphism of $A \ltimes DA$-bimodules. Let $(a, f), (a', f'), (b, g) \in A \ltimes DA$; then

$$\phi((a, f) + (a', f'))(b, g) = (f + f')(b) + g(a + a')$$

is equal to

$$\phi(a, f)(b, g) + \phi(a', f')(b, g) = f(b) + g(a) + f'(b) + g(a'),$$

which shows that ϕ is a morphism of additive groups.

To show that ϕ preserves the right $A \ltimes DA$-action, we compute

$$\begin{aligned}
\phi\big((a, f)(a', f')\big)(b, g) &= \phi(aa', fa' + af')(b, g) \\
&= (fa')(b) + (af')(b) + g(aa') \\
&= f(a'b) + f'(ba) + g(aa') \\
&= f(a'b) + (f'b)(a) + (a'g)(a) \\
&= \phi(a, f)(a'b, f'b + a'g) \\
&= \big(\phi(a, f)\big)(a', f')(b, g).
\end{aligned}$$

In a similar computation, one can show that ϕ also preserves the left $A \ltimes DA$-action. So ϕ is a morphism of $A \ltimes DA$-bimodules.

Next, we show that ϕ is injective. Suppose that $(a, f) \in \ker \phi$. Then $f(b) + g(a) = 0$, for all $(b, g) \in A \ltimes DA$. If we take $g = 0$, we see that $f(b) = 0$ for all $b \in A$, thus $f = 0$. But if $f = 0$, then $g(a) = 0$ for all $g \in DA$, and therefore $a = 0$. Thus $(a, f) = (0, 0)$, and ϕ is injective.

The surjectivity of ϕ now follows simply from the fact that the source and the target of ϕ are finite-dimensional vector spaces of the same dimension. □

Example 6.6. Let A be the path algebra of the quiver

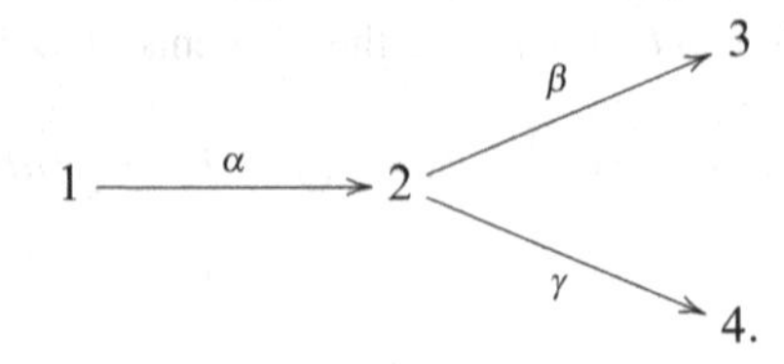

Then the A-modules A and DA have the following direct sum decompositions:

$$A \quad \begin{smallmatrix} 1 \\ 2 \\ 3\,4 \end{smallmatrix} \oplus \begin{smallmatrix} 2 \\ 3\,4 \end{smallmatrix} \oplus 3 \oplus 4 ,$$

$$DA \quad 1 \oplus \begin{smallmatrix} 1 \\ 2 \end{smallmatrix} \oplus \begin{smallmatrix} 1 \\ 2 \\ 3 \end{smallmatrix} \oplus \begin{smallmatrix} 1 \\ 2 \\ 4 \end{smallmatrix} .$$

Now we want to construct the indecomposable projective modules $P(i)$ over the trivial extension algebra $A \ltimes DA$. By Proposition 6.1 (6), there is a short exact sequence:

$$0 \longrightarrow I_A(i) \longrightarrow P(i) \longrightarrow P_A(i) \longrightarrow 0 ,$$

where $I_A(i)$ and $P_A(i)$ denote the injective and the projective A-module at vertex i considered as $A \ltimes DA$-modules. Therefore the quiver of $A \ltimes DA$ is

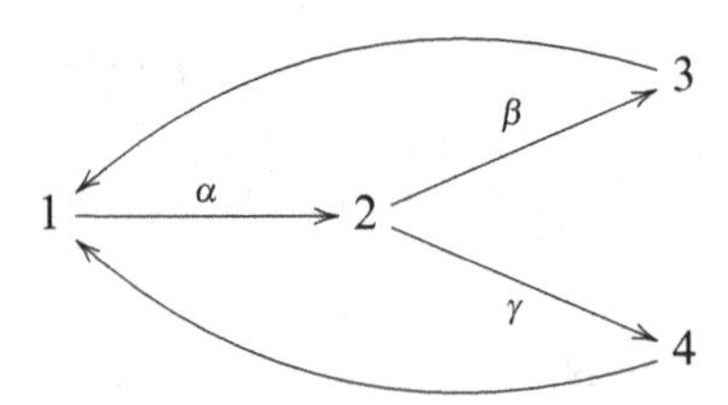

and the $A \ltimes DA$-module $A \ltimes DA$ is the following direct sum of indecomposable projective–injective modules:

$$A \ltimes DA \quad \begin{smallmatrix} 1 \\ 2 \\ 3\,4 \\ 1 \end{smallmatrix} \oplus \begin{smallmatrix} 2 \\ 3\,4 \\ 1 \\ 2 \end{smallmatrix} \oplus \begin{smallmatrix} 3 \\ 1 \\ 2 \\ 3 \end{smallmatrix} \oplus \begin{smallmatrix} 4 \\ 1 \\ 2 \\ 4 \end{smallmatrix}$$

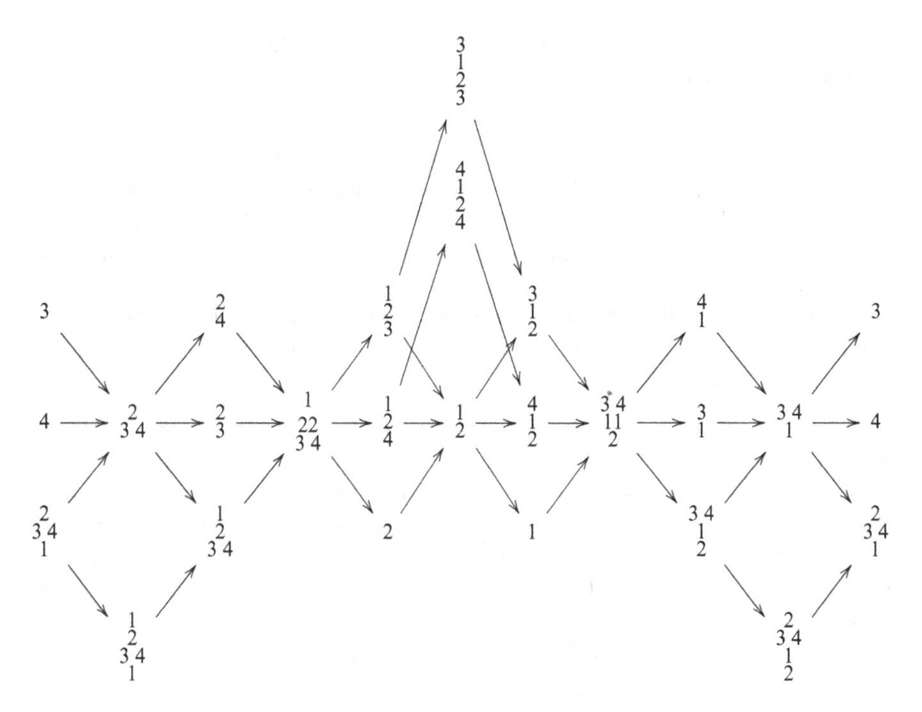

Fig. 6.1 Auslander–Reiten quiver of $A \ltimes DA$ in Example 6.6

The Auslander–Reiten quiver of $A \ltimes DA$ is illustrated in Fig. 6.1; the modules that have the same label have to be identified.

6.4 Cluster-Tilted Algebras

In this subsection, we study the trivial extensions $C \ltimes M$ of Sect. 6.2 in the special case where C is a tilted algebra and M is the C-bimodule $\mathrm{Ext}^2_C(DC, C)$. The C-bimodule structure on $\mathrm{Ext}^2_C(DC, C)$ is given by

$$a\left(\mathrm{Ext}^2_C(DC, C)\right) b = \mathrm{Ext}^2_C(D(Cb), aC).$$

Definition 6.5. Let C be a tilted algebra of type Q. Then the trivial extension

$$C \ltimes \mathrm{Ext}^2_C(DC, C)$$

is called a **cluster-tilted algebra** of type Q.

Remark 6.6. If the tilted algebra C is hereditary, then the bimodule $\mathrm{Ext}^2_C(DC, C)$ is zero, and the cluster-tilted algebra is equal to C. Thus *hereditary algebras are cluster-tilted algebras*.

Remark 6.7. If the tilted algebra C is of type $\mathbb{A}_n$, then the cluster-tilted algebra $C \ltimes \mathrm{Ext}^2_C(DC, C)$ is a bound quiver algebra associated to a triangulation of the polygon with $n + 3$ vertices as in Sect. 3.4.1. The cluster-tilted algebra is hereditary if and only if, in the corresponding triangulation, every triangle has at least one side on the boundary of the polygon.

Similarly, if the tilted algebra is of type $\mathbb{D}_n$, then the cluster-tilted algebra is a bound quiver algebra associated to a triangulation of the punctured polygon with n vertices as in Sect. 3.4.2.

Let us write $\tilde{C}$ for the cluster-tilted algebra of C. Let Q be the quiver of the tilted algebra C and let $Q_0 = \{1, 2, \ldots, n\}$ be the set of vertices. We want to describe the quiver $\tilde{Q}$ of $\tilde{C}$. It follows from Proposition 6.1 that $\mathrm{Ext}^2_C(DC, C)$ is contained in the radical of $\tilde{C}$, that $1_{\tilde{C}} = (1_C, 0)$, and that the primitive orthogonal idempotents $e_1, \ldots, e_n \in C$ given by the constant paths remain primitive orthogonal idempotents in $\tilde{C}$ and $1_{\tilde{C}} = e_1 + \cdots + e_n$. It follows that the quiver $\tilde{Q}$ has the same set of vertices as the quiver Q.

The arrows of $\tilde{Q}$ correspond to a basis of the vector space:

$$\mathrm{rad}\,\tilde{C}/\mathrm{rad}^2\tilde{C}.$$

Again by Proposition 6.1, we know that the vector space $\mathrm{rad}\,\tilde{C}$ has a direct sum structure $\mathrm{rad}\,\tilde{C} = \mathrm{rad}\,C \oplus \mathrm{Ext}^2_C(DC, C)$ and that the square of $\mathrm{Ext}^2_C(DC, C)$ is zero. Therefore

$$\mathrm{rad}^2\tilde{C} = \mathrm{rad}^2 C + \mathrm{rad}\,C\,\mathrm{Ext}^2_C(DC, C) + \mathrm{Ext}^2_C(DC, C)\mathrm{rad}\,C,$$

and, hence, $\mathrm{rad}\,\tilde{C}/\mathrm{rad}^2\tilde{C}$ is equal to

$$\frac{\mathrm{rad}\,C}{\mathrm{rad}^2 C} \oplus \frac{\mathrm{Ext}^2_C(DC, C)}{(\mathrm{rad}\,C\,\mathrm{Ext}^2_C(DC, C) + \mathrm{Ext}^2_C(DC, C)\mathrm{rad}\,C)}.$$

Therefore there are two types of arrows in $\tilde{Q}$:

- the arrows corresponding to a basis of $\mathrm{rad}\,C/\mathrm{rad}^2 C$; these are the arrows in the quiver Q, and we call them the *old arrows*;
- the arrows corresponding to a basis of

  $$\mathrm{Ext}^2_C(DC, C)/(\mathrm{rad}\,C\,\mathrm{Ext}^2_C(DC, C) + \mathrm{Ext}^2_C(DC, C)\mathrm{rad}\,C),$$

  which we call the *new arrows*.

A *new path* in $\tilde{Q}$ is a path that contains a new arrow. The nonzero new paths correspond to a basis of $\mathrm{Ext}^2_C(DC, C)$ and a nonzero new path from i to j is an element of $\mathrm{Ext}^2_C(I(j), P(i))$. Note that each nonzero new path contains *exactly* one of the new arrows because the square of the ideal $\mathrm{Ext}^2_C(DC, C)$ is zero in $\tilde{C}$, by Proposition 6.1.

Example 6.7. Let C be the tilted algebra of type $\mathbb{D}_4$ of Example 6.5 given by the quiver

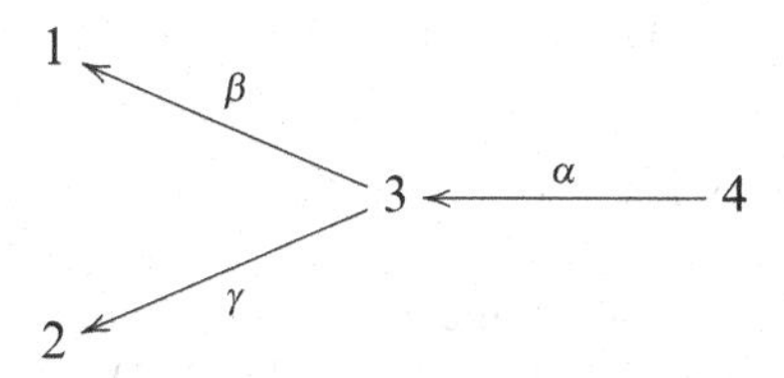

and relations $\alpha\beta = 0$, $\alpha\gamma = 0$. Then the indecomposable projective, and the indecomposable injective C-modules are as follows:

$$P(1) = 1 \quad P(2) = 2 \quad P(3) = \begin{smallmatrix} 3 \\ 1\,2 \end{smallmatrix} \quad P(4) = \begin{smallmatrix} 4 \\ 3 \end{smallmatrix}$$

$$I(1) = \begin{smallmatrix} 3 \\ 1 \end{smallmatrix} \quad I(2) = \begin{smallmatrix} 3 \\ 2 \end{smallmatrix} \quad I(3) = \begin{smallmatrix} 4 \\ 3 \end{smallmatrix} \quad I(4) = 4$$

To compute $\mathrm{Ext}^2_C(DC, C)$, we construct a minimal projective resolution for each indecomposable injective. For $I(1)$, we get

$$0 \longrightarrow P(2) \longrightarrow P(3) \longrightarrow I(1) \longrightarrow 0,$$

which shows that the projective dimension of $I(1)$ is equal to 1, and therefore $\mathrm{Ext}^2_C(I(1), C) = 0$.

A similar computation shows that $\mathrm{Ext}^2_C(I(2), C) = 0$.

For $I(3)$, we get

$$0 \longrightarrow P(4) \longrightarrow I(3) \longrightarrow 0,$$

so the projective dimension of $I(3)$ is 0 and again $\mathrm{Ext}^2_C(I(3), C) = 0$.

Finally, for $I(4)$, we get

$$0 \xrightarrow{g} P(1) \oplus P(2) \xrightarrow{f} P(3) \longrightarrow P(4) \longrightarrow I(4) \longrightarrow 0,$$

thus the projective dimension of $I(4)$ is 2, and therefore $\mathrm{Ext}^2_C(I(4), C)$ is a nonzero vector space. To compute a basis of it, we apply the functor $\mathrm{Hom}(-, C)$ to the projective resolution and obtain the complex:

$$0 \longrightarrow \operatorname{Hom}(I(4), C) \longrightarrow \operatorname{Hom}(P(4), C) \longrightarrow$$

$$\longrightarrow \operatorname{Hom}(P(3), C) \xrightarrow{f^*} \operatorname{Hom}(P(1) \oplus P(2), C) \xrightarrow{g^*} 0.$$

By definition, $\operatorname{Ext}^2(I(4), C)$ is equal to the quotient $\ker g^*/\operatorname{im} f^*$, and since g^* is the zero map, we have

$$\operatorname{Ext}^2(I(4), C) = \operatorname{Hom}(P(1) \oplus P(2), C)/\operatorname{im} f^*.$$

The space $\operatorname{Hom}(P(1) \oplus P(2), C)$ is the direct sum of all $\operatorname{Hom}(P(1), P(j))$ and all $\operatorname{Hom}(P(2), P(j))$, $j = 1, 2, 3, 4$. Let us consider $P(1)$ first. Since $P(2)$ and $P(4)$ are of dimension zero at the vertex 1, we see that $\operatorname{Hom}(P(1), P(j))$ is zero if $j = 2$ or 4; and, since $P(1)$ and $P(3)$ are of dimension one at the vertex 1, we see that $\operatorname{Hom}(P(1), P(j))$ is one-dimensional if $j = 1$ or 3.

Let f_1 denote the restriction of f to $P(1)$. Since f_1 is a nonzero morphism in $\operatorname{Hom}(P(1), P(3))$, we see that any morphism $h \in \operatorname{Hom}(P(1), P(3))$ is a scalar multiple of f_1; thus $h = \lambda f_1$ for some $\lambda \in k$. This implies that $h = \lambda 1_{P(3)} \circ f_1 = f_1^*(\lambda 1_{P(3)})$, and thus $h \in \operatorname{im} f^*$, and, therefore, the class of h in $\operatorname{Ext}^2(I(4), C)$ is zero.

On the other hand, $\operatorname{Hom}(P(1), P(1))$ is the one-dimensional vector space with basis the identity map $1_{P(1)}$. Since there is no nonzero morphism from $P(3)$ to $P(1)$, we see that the $1_{P(1)}$ does not factor through f_1, and therefore the class of $1_{P(1)}$ in $\operatorname{Ext}^2(I(4), C)$ is nonzero.

This shows that $\operatorname{Ext}^2(I(4), P(1))$ is one-dimensional with basis the class of $1_{P(1)}$. In a similar way, one can show that $\operatorname{Ext}^2(I(4), P(2))$ is one-dimensional with basis the class of $1_{P(2)}$, and we conclude that $\operatorname{Ext}^2(DC, C)$ is of dimension two and decomposes as

$$\operatorname{Ext}^2(DC, C) = \operatorname{Ext}^2(I(4), P(1)) \oplus \operatorname{Ext}^2(I(4), P(2)).$$

This shows that, in this example, there are exactly two new paths: one from 1 to 4 and the other from 2 to 4. Both of them are given by a new arrow, so the quiver of the cluster-tilted algebra is

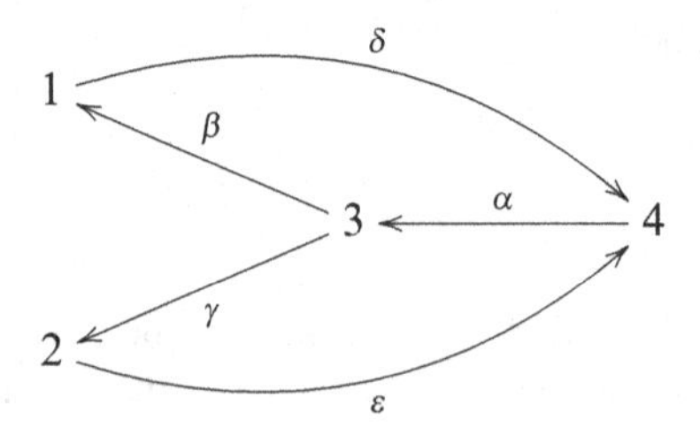

with relations $\alpha\beta = 0$, $\alpha\gamma = 0$, $\delta\alpha = 0$, $\epsilon\alpha = 0$, $\beta\delta = \gamma\epsilon$.

Remark 6.8. In the example above, the quiver $\tilde{Q}$ is obtained from the quiver of Q by adding one arrow for each of the relations we used to define the algebra C: the relation $\alpha\beta$ gives rise to the arrow δ, and the relation $\alpha\gamma$ gives rise to the arrow ϵ. The new arrow goes from the terminal point to the initial point of the relation. So the relation together with the new arrow forms an oriented cycle.

This is also true in general, that is, given a minimal system of relations $\rho_1, \ldots, \rho_t$ such that the tilted algebra $C = kQ/\langle\rho_1, \ldots, \rho_t\rangle$, then the quiver $\tilde{Q}$ of the cluster-tilted algebra is obtained from Q by adding the arrows $\alpha_i : t(\rho_i) \to s(\rho_i)$. For a proof of this result see [4].

6.5 Triangular Matrix Algebras

Suppose that the algebra A is a direct product of algebras $A = A_1 \times A_2$, and thus $A = A_1 \oplus A_2$ as vector spaces, and the multiplication is given componentwise $(a_1, a_2)(a_1', a_2') = (a_1 a_1', a_2 a_2')$. Suppose further that the bimodule structure of M is as follows:

$$(a_1, a_2)m = a_2 m \qquad \text{and} \qquad m(a_1, a_2) = m a_1,$$

for all $a_1 \in A_1, a_2 \in A_2$ and $m \in M$. In other words, M is an (A_2, A_1)-bimodule. Then the product in $A \ltimes M$ becomes

$$\big((a_1, a_2), m\big)\big((a_1', a_2'), m'\big) = \big((a_1 a_1', a_2 a_2'), a_2 m' + m a_1'\big).$$

In this situation, it is convenient to represent the elements of $A \ltimes M$ as 2×2 matrices $\begin{bmatrix} a_1 & 0 \\ m & a_2 \end{bmatrix}$, because then the product becomes the ordinary matrix multiplication:

$$\begin{bmatrix} a_1 & 0 \\ m & a_2 \end{bmatrix}\begin{bmatrix} a_1' & 0 \\ m' & a_2' \end{bmatrix} = \begin{bmatrix} a_1 a_1' & 0 \\ m a_1' + a_2 m' & a_2 a_2' \end{bmatrix}. \tag{6.2}$$

This leads us to the following definition:

Definition 6.6. Let A_1 and A_2 be two algebras and M an (A_2, A_1)-bimodule. Then the algebra

$$\begin{bmatrix} A_1 & 0 \\ M & A_2 \end{bmatrix} = \left\{ \begin{bmatrix} a_1 & 0 \\ m & a_2 \end{bmatrix} \middle| \, a_1 \in A_1, a_2 \in A_2, m \in M \right\},$$

with product given by equation (6.2), is called **triangular matrix algebra**.

Proposition 6.9. *Let A be the triangular matrix algebra*

$$\begin{bmatrix} A_1 & 0 \\ M & A_2 \end{bmatrix}.$$

Then

1.

$$1_A = \begin{bmatrix} 1_{A_1} & 0 \\ 0 & 1_{A_2} \end{bmatrix}$$

2. $M = \begin{bmatrix} 0 & 0 \\ M & 0 \end{bmatrix}$ *is a two-sided nilpotent ideal and* $M^2 = 0$.

3.

$$\text{rad}\, A = \begin{bmatrix} \text{rad}\, A_1 & 0 \\ M & \text{rad}\, A_2 \end{bmatrix}$$

4. *Given primitive orthogonal idempotents $e_i \in A_1$ and $f_i \in A_2$ such that $1_{A_1} = e_1 + \ldots + e_{n_1}$ and $1_{A_2} = f_1 + \ldots + f_{n_2}$, then the indecomposable projective A-modules are of the form*

$$\begin{bmatrix} e_i A_1 & 0 \\ 0 & 0 \end{bmatrix} \quad \text{or} \quad \begin{bmatrix} 0 & 0 \\ f_i M & f_i A_2 \end{bmatrix}.$$

Proof. (1)–(3) follow directly from Proposition 6.1. Using (1), we see that $1_A = e_1 + \cdots + e_{n_1} + f_1 + \cdots + f_{n_2}$ and therefore (4) follows by multiplying the algebra with the idempotents $\begin{bmatrix} e_i & 0 \\ 0 & 0 \end{bmatrix}$ and $\begin{bmatrix} 0 & 0 \\ 0 & f_i \end{bmatrix}$ on the left. □

6.6 One-Point Extensions

Now we specialize our construction of Sect. 6.5 even further, taking A_2 to be the field k and M a right A_1-module.

Definition 6.7. Let B be a k-algebra and M a right B-module. The triangular matrix algebra

$$B[M] = \begin{bmatrix} B & 0 \\ M & k \end{bmatrix}$$

is called the **one-point extension** of B by the module M.

The terminology comes from the fact that if B is given as a bound quiver algebra, then the quiver of $B[M]$ contains the quiver of B and has exactly one vertex more.

Example 6.8. Let B be given by the quiver

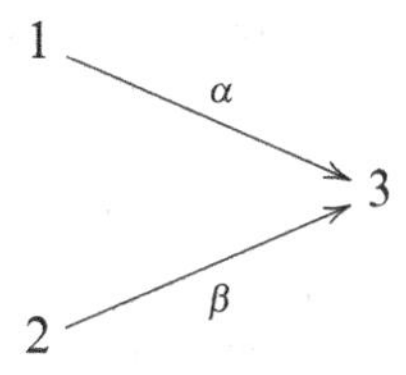

and let $M = S(1) \oplus S(2)$. Then

$$B = \begin{bmatrix} k & 0 & k \\ 0 & k & k \\ 0 & 0 & k \end{bmatrix} \quad \text{and} \quad B[M] = \begin{bmatrix} k & 0 & k & 0 \\ 0 & k & k & 0 \\ 0 & 0 & k & 0 \\ k & k & 0 & k \end{bmatrix}.$$

Thus $B[M]$ is given by the quiver

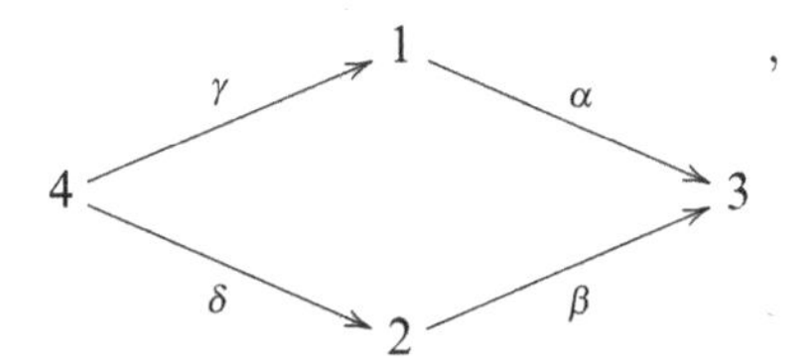

with relations $\gamma\alpha = 0 = \delta\beta$.

Note that the radical of the new projective $P(4)$ is exactly the module M.

Proposition 6.10. *Let B be the bound quiver algebra kQ_B/I and let $B[M]$ be the one-point extension of B by the B-module M. Then the quiver of $B[M]$ is obtained from the quiver of B by adding one vertex w and*

1. *For all vertices $i \neq w$, the indecomposable projective $B[M]$-module $P(i)$ is isomorphic to the indecomposable projective B-module $P(i)$.*
2. *w is a source in the quiver of $B[M]$.*
3. $\operatorname{rad} B[M] = \begin{bmatrix} \operatorname{rad} B & 0 \\ M & 0 \end{bmatrix}$.
4. $\operatorname{rad}_{B[M]} P(i) \cong \begin{cases} \operatorname{rad}_B P(i) & \text{if } i \text{ is a vertex of } Q_B, \\ M & \text{if } i \text{ is the new vertex } w. \end{cases}$

Proof. (1),(3), and (4) follow from Proposition 6.9. Moreover, (1) implies that there are no paths from a vertex $i \neq w$ to w, and this shows (2). □

6.7 Notes

The description of algebras using quivers and relations goes back to Gabriel [33,34].

Tilted algebras were introduced in [22, 39]. For an introduction to tilting theory, we refer to [8,52] and further information and generalizations can be found in [2,9, 23,29,36–38,51].

For trivial extensions in general, we refer to [32] and for the trivial extensions $A \ltimes DA$ to [7,40–42,57].

Cluster-tilted algebras were introduced in [25] and [28] and studied in [3–6, 16, 19,20,24,26,27,44,48,50,58], and many others.

Problems

Exercises for Chapter 6

6.1. Find all tilted algebras of type $\mathbb{A}_4$.

6.2. Find all tilted algebras of type $\mathbb{D}_5$.

6.3. Show that the algebra in Example 5.14 is cluster-tilted.

6.4. Let C be the algebra given by the quiver

$$1 \xrightarrow{\alpha} 2 \xrightarrow{\beta} 3 \xrightarrow{\gamma} 4$$

bound by the relation $\alpha\beta\gamma$. This algebra is tilted of type $\mathbb{D}_4$. Compute $\mathrm{Ext}^2_C(DC, C)$ and construct the quiver of the corresponding cluster-tilted algebra.

6.5. Let C be the bound quiver algebra given by the quiver

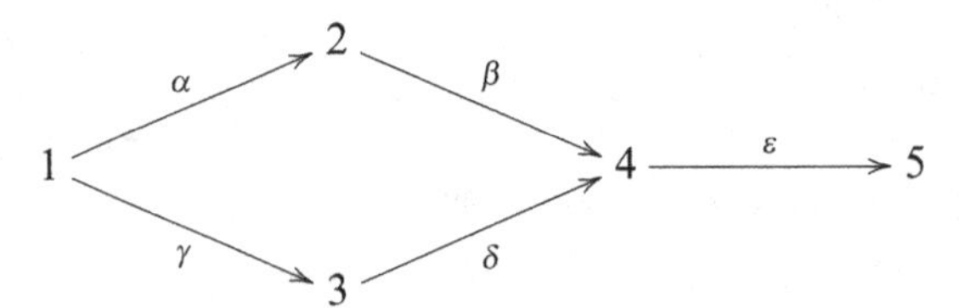

and relations $\alpha\beta - \gamma\delta, \alpha\beta\epsilon$. This algebra is tilted of type $\tilde{\mathbb{D}}_4$ (do not show this). Compute $\mathrm{Ext}^2_C(DC, C)$.

6.6. Let $B = A \ltimes M$ be the trivial extension algebra of the algebra A with respect to the bimodule M. Let N be a right A-module and define a right B-action on N by

$$\begin{array}{rcl} N \times & B & \longrightarrow N \\ (n \quad , & (a,m)) & \longmapsto na \end{array}$$

Show that N is a right B-module with respect to this action.

6.7. Let A be the path algebra of the quiver

$$1 \longrightarrow 2 \longrightarrow 3\,.$$

Compute the Auslander–Reiten quiver of the trivial extension:

$$A \ltimes DA.$$

6.8. Show that the path algebra kQ of the quiver

$$Q = \; 1 \overset{\alpha}{\underset{\beta}{\rightrightarrows}} 2$$

is isomorphic to a one-point extension $k[M]$ of the algebra k with respect to the module $M = k \oplus k$.

6.9. Let $B = kQ$ be the path algebra of the quiver

$$Q = \; 1 \overset{\alpha}{\underset{\beta}{\rightrightarrows}} 2$$

and let M_λ be the representation

$$k \overset{1}{\underset{\lambda}{\rightrightarrows}} k\,.$$

Show that the one-point extension $B[M_\lambda]$ is isomorphic to the bound quiver algebra given by

$$3 \xrightarrow{\;\gamma\;} 1 \overset{\alpha}{\underset{\beta}{\rightrightarrows}} 2\,,$$

bound by the relation $\lambda\gamma\alpha = \gamma\beta$.

Chapter 7
Auslander–Reiten Theory

Recall that the goal of representation theory is to classify the indecomposable modules and the morphisms between them. The Auslander–Reiten quiver is a first approximation of the module category. If the quiver is of finite representation type, then the Auslander–Reiten quiver gives a complete picture of the module category.

In earlier chapters, we have seen several examples of Auslander–Reiten quivers. In this chapter, we develop the theoretical background for them.

Recall that the vertices of the Auslander–Reiten quiver of an algebra A correspond to the isoclasses of indecomposable A-modules, and its arrows correspond to the irreducible morphisms between indecomposable A-modules. In addition to its vertices and arrows the Auslander–Reiten quiver has an extra structure, the Auslander–Reiten translation τ. We have seen in examples throughout the book how τ is used to construct the meshes of the Auslander–Reiten quiver. Each such mesh corresponds to a short exact sequence of the form

$$0 \longrightarrow \tau M \longrightarrow E \longrightarrow M \longrightarrow 0\,,$$

called almost split sequence. By definition, an almost split sequence is a short exact sequence

$$0 \longrightarrow L \xrightarrow{f} M \xrightarrow{g} N \longrightarrow 0$$

whose morphisms f and g satisfy a certain, rather technical, condition. We study this definition and its relation to irreducible morphisms in the first section.

Another way of thinking about an almost split sequence is as a canonical extension between a module and its Auslander–Reiten translate. Back in Chap. 3 when we constructed examples of Auslander–Reiten quivers of types $\mathbb{A}$ and $\mathbb{D}$, we computed the dimensions of Hom-spaces and Ext-spaces in a combinatorial way from the Auslander–Reiten quivers. In these examples, we observed that $\dim \operatorname{Ext}^1(M, N) = \dim \operatorname{Hom}(N, \tau M)$ for any two modules M and N. This result is a direct consequence of the Auslander–Reiten formulas (Theorem 7.18) which state that

R. Schiffler, *Quiver Representations*, CMS Books in Mathematics,
DOI 10.1007/978-3-319-09204-1_7

$$\mathrm{Ext}^1(M,N) \cong D\underline{\mathrm{Hom}}(\tau^{-1}N,M) \cong D\overline{\mathrm{Hom}}(N,\tau M),$$

where the overlining, respectively underlining, means that we are considering morphisms modulo factorization through projective modules, respectively injective modules.

The Auslander–Reiten formulas are a fascinating result relating the exact structure of the module category to its morphism structure. Every short exact sequence corresponds to a morphism, and the composition of morphisms corresponds to a composition of short exact sequences. Now, the most elementary morphisms in the module category are the identity morphisms $1_M \in \mathrm{Hom}(M,M)$. If M is indecomposable and non-projective, then the Auslander–Reiten formulas imply that

$$D\overline{\mathrm{Hom}}(M,M) \cong \mathrm{Ext}^1(M,\tau M),$$

and therefore the image of (the dual of) the identity morphism 1_M under this isomorphism is a canonical element of $\mathrm{Ext}^1(M,\tau M)$, thus a short exact sequence of the form

$$0 \longrightarrow \tau M \longrightarrow E \longrightarrow M \longrightarrow 0\,.$$

This sequence is an almost split sequence.

7.1 Almost Split Sequences

Let A be a finite-dimensional k-algebra.

From the examples that we have already seen, we know that the almost split sequences are the meshes of the Auslander–Reiten quiver. One might therefore expect that an almost split sequence is a non-split short exact sequence of the form $0 \to \tau M \to E \to M \to 0$. We will see later that every almost split sequence is of this form, but, unfortunately, one cannot use this property as a definition of almost split sequences, because not every such sequence is almost split.

For example, consider the Kronecker quiver $1 \rightrightarrows 2$ and the representations

$$M = k^2 \underset{\begin{bmatrix}0&1\\0&0\end{bmatrix}}{\overset{\begin{bmatrix}1&0\\0&1\end{bmatrix}}{\rightrightarrows}} k^2 \quad\text{and}\quad E = k^4 \underset{\begin{bmatrix}0&1&0&0\\0&0&1&0\\0&0&0&1\\0&0&0&0\end{bmatrix}}{\overset{\begin{bmatrix}1&0&0&0\\0&1&0&0\\0&0&1&0\\0&0&0&1\end{bmatrix}}{\rightrightarrows}} k^4\,.$$

Then one can show that $\tau M = M$ and there is a non-split short exact sequence

$$0 \longrightarrow \tau M \longrightarrow E \longrightarrow M \longrightarrow 0$$

which is not almost split. The almost split sequence in this example would be

$$0 \longrightarrow \tau M \longrightarrow E_1 \oplus E_2 \longrightarrow M \longrightarrow 0$$

with

$$E_1 = k^3 \overset{\begin{bmatrix} 1&0&0\\0&1&0\\0&0&1\end{bmatrix}}{\underset{\begin{bmatrix} 0&1&0\\0&0&1\\0&0&0\end{bmatrix}}{\rightrightarrows}} k^3 \quad \text{and} \quad E_2 = k \overset{1}{\underset{0}{\rightrightarrows}} k\,.$$

In order to define the almost split sequences, we first need to introduce the concept of almost split morphisms.

Definition 7.1. A morphism $f : L \to M$ is called **left minimal almost split** if

1. f is not a section, that is, there is no morphism $h : M \to L$ such that $hf = 1_L$;
2. for each morphism $u : L \to U$ in mod A which is not a section, there exists a morphism $u' : M \to U$ such that $u'f = u$;

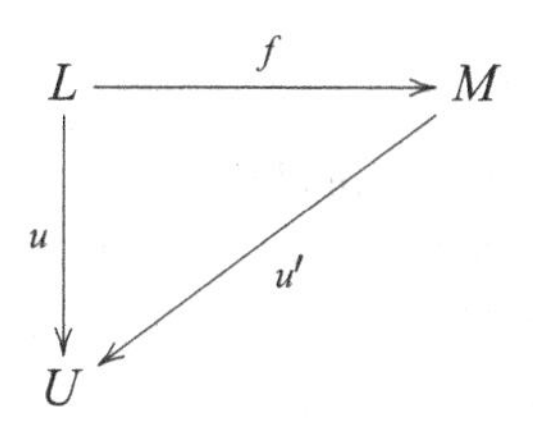

3. if $h : M \to M$ is such that $hf = f$ then h is an automorphism of M.

Similarly, a morphism $g : M \to N$ is called **right minimal almost split** if

(1') g is not a retraction, that is, there is no morphism $h : N \to M$ such that $gh = 1_N$;
(2') for each morphism $v : V \to N$ in mod A which is not a retraction, there exists a morphism $v' : V \to M$ such that $gv' = v$:

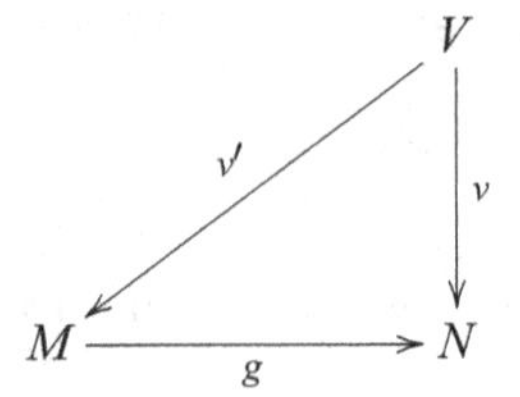

(3') if $h : M \to M$ is such that $gh = g$, then h is an automorphism of M.

Now we are ready to define almost split sequences.

Definition 7.2. A short exact sequence in mod A

$$0 \longrightarrow L \xrightarrow{f} M \xrightarrow{g} N \longrightarrow 0$$

is called an **almost split sequence** (or Auslander–Reiten sequence) if f is a left minimal almost split morphism and g is a right minimal almost split morphism.

Remark 7.1. An almost split sequence is not split by conditions (1) and (1').

Remark 7.2. In condition (2) of Definition 7.1, we require that u is not a section because otherwise there would be a morphism h such that $hu = 1_L$ and thus $(hu')f = hu = 1_L$ and f would be a section too:

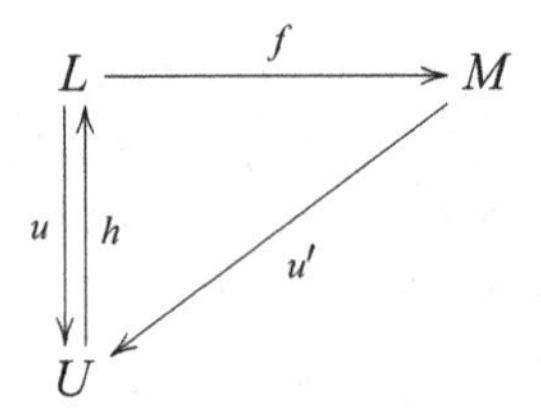

There is an equivalent definition of almost split sequences in terms of irreducible morphisms that we will also present here. The irreducible morphisms correspond to the arrows in the Auslander–Reiten quiver.

Definition 7.3. A morphism $f : X \to Y$ in mod A is called **irreducible** if

- f is not a section;
- f is not a retraction;
- and whenever $f = gh$ for some morphisms $h : X \to Z$ and $g : Z \to Y$, then either h is a section or g is a retraction.

One can characterize almost split sequences by irreducible morphism.

Lemma 7.3. *A short exact sequence in* mod A

$$0 \longrightarrow L \xrightarrow{f} M \xrightarrow{g} N \longrightarrow 0$$

is an almost split sequence if and only if L and N are indecomposable modules and f and g are irreducible morphisms. Moreover, in this situation, we have $L \cong \tau N$.

Proof. See [14, V.5.9 and V.1.14] or [8, IV.1.13]. □

Proposition 7.4. *Let $f : X \to Y$ be an irreducible morphism. Then*

1. *f is either injective or surjective (but not both),*
2. *f admits no nontrivial factorization.*

Proof.

(1) Suppose that f is not surjective and define the morphism $\overline{f} : X \to \operatorname{im} f$ by $\overline{f}(x) = f(x)$, and let $i : \operatorname{im} f \to Y$ be the inclusion morphism. Since f is not surjective, we see that i is not a retraction. Then $f = i\overline{f}$ and, f being irreducible, it follows that $\overline{f}$ is a section, hence injective. Consequently, f is injective too.

Note that f cannot be both injective and surjective, since f is not a section.

(2) Consider a factorization of f, that is, a morphism $g : X \to Z$ and a morphism $h : Z \to Y$ such that $f = hg$. Since f is irreducible, it follows that g is a section or h is a retraction.

If f is injective, then f is not surjective, and thus h cannot be a retraction. Then g must be a section, and $X \cong \operatorname{im} g$ is a direct summand of Z, which shows that the factorization is trivial.

On the other hand, if f is not injective, then f is surjective by (1). Then a similar argument shows that h is a retraction, Y is isomorphic to a direct summand Y' of Z, and that $Z = Y' \oplus \ker h$. Again it follows that the factorization is trivial. □

Our next goal is to define the Auslander–Reiten quiver. We know already that its arrows $M \to N$ correspond to irreducible morphisms in $\operatorname{Hom}(M, N)$, but we need to make precise how many arrows we draw between M and N, and for that we need a way of saying how big the space of irreducible morphism morphisms from M to N is.

For example, let A be the path algebra of the Kronecker quiver

$$1 \underset{\beta}{\overset{\alpha}{\leftleftarrows}} 2$$

and consider the A-modules $P(1) = 1$ and $P(2) = \begin{smallmatrix} & 2 & \\ 1 & & 1 \end{smallmatrix}$. Then every morphism from $P(1)$ to $P(2)$ is irreducible, so the dimension of the space of irreducible morphisms from $P(1)$ to $P(2)$ is the same as the dimension of $\operatorname{Hom}(P(1), P(2))$, which is equal to 2. Hence, in the Auslander–Reiten quiver, we draw 2 arrows from $P(1)$ to $P(2)$.

To make these ideas precise, we introduce one more notion of category theory.

Categories 7 *The notion of an ideal in a ring can be generalized to the notion of an ideal in an additive category $\mathscr{C}$. We will define here the radical of the category, which is an example of such an ideal.*

Definition 7.4. *Let $\mathscr{C}$ be an additive k-category. An* ideal *I in $\mathscr{C}$ is a class of morphisms in $\mathscr{C}$ such that for all objects X, Y, Z in $\mathscr{C}$ and all morphisms $f, f' \in$ $\mathrm{Hom}(X, Y)$, $g \in \mathrm{Hom}(Y, Z)$ we have*

1. *the zero morphism $X \to X, x \mapsto 0$ is in I,*
2. *if $f, f' \in I$ and $\lambda, \lambda' \in k$, then $\lambda f + \lambda' f' \in I$, thus I is closed under addition and scalar multiplication,*
3. *a. if $f \in I$, then $g \circ f \in I$,*
   b. if $g \in I$, then $g \circ f \in I$,
   thus I is stable under the composition on the left and on the right.

If I is an ideal in $\mathscr{C}$, then let $I(X, Y)$ denote the class of morphisms $X \to Y$ in I. Note that if I contains a section $f : X \to Y$ with $h : Y \to X$ such that $hf = 1_X$, then $1_X \in I$, and then I contains any morphism starting or ending at X; thus $\mathrm{Hom}_{\mathscr{C}}(X, Z)$ and $\mathrm{Hom}_{\mathscr{C}}(Z, X)$ lie in I, for all objects Z in $\mathscr{C}$. Similarly, if I contains a retraction $f : X \to Y$, then $\mathrm{Hom}_{\mathscr{C}}(Y, Z)$ and $\mathrm{Hom}_{\mathscr{C}}(Z, Y)$ lie in I, for all objects Z.

This shows in particular that, if an ideal contains for every object X a section $f : X \to Y$ or a retraction $f : Y \to X$, then the ideal contains all morphisms in the category.

We now define the radical of a category. Recall that for a ring R, we have shown in Lemma 4.1 that $a \in \mathrm{rad}\, R$ if and only if $1 - ab$ is invertible for all $b \in R$. We use this description of the radical of a ring to *define* the radical of a category.

Definition 7.5. The (Jacobson) **radical** $\mathrm{rad}_{\mathscr{C}}$ of the additive k-category $\mathscr{C}$ is defined to be the class of all morphisms such that for any pair of objects X, Y in $\mathscr{C}$, we have that $\mathrm{rad}_{\mathscr{C}}(X, Y)$ is equal to

$$\{f \in \mathrm{Hom}_{\mathscr{C}}(X, Y) \mid (1_X - hf) \text{ is an iso for all } h \in \mathrm{Hom}_{\mathscr{C}}(Y, X)\}.$$

Note that if $f \in \mathrm{rad}_{\mathscr{C}}(X, Y)$, then f is not an isomorphism, because otherwise $1_X - f^{-1} f = 0$.

Lemma 7.5. $\mathrm{rad}_{\mathscr{C}}$ *is an ideal in $\mathscr{C}$.*

Proof. Clearly $\mathrm{rad}_{\mathscr{C}}$ satisfies condition (1) of Definition 7.4. To show condition (2), let $f, f' \in \mathrm{rad}_{\mathscr{C}}(X, Y)$, $\lambda, \lambda' \in k$, and $h \in \mathrm{Hom}_{\mathscr{C}}(Y, X)$. Let $b \in \mathrm{Hom}_{\mathscr{C}}(X, X)$ be such that

$$b(1_X - h\lambda f) = 1_X.$$

Note that b is an isomorphism. Then $b(1_X - h(\lambda f + \lambda' f')) = 1_X - bh\lambda' f'$ which is an isomorphism since $f' \in \mathrm{rad}_{\mathscr{C}}$. It follows that $(1_X - h(\lambda f + \lambda' f'))$ is an isomorphism, and thus $\lambda f + \lambda' f' \in \mathrm{rad}_{\mathscr{C}}$. This shows condition (2).

Finally, if $f \in \mathrm{rad}_{\mathscr{C}}(X, Y)$ and $g \in \mathrm{Hom}_{\mathscr{C}}(Y, Z)$, then for every $h \in \mathrm{Hom}_{\mathscr{C}}(Z, X)$, we have $hg \in \mathrm{Hom}_{\mathscr{C}}(Y, X)$, and thus $1_X - (hg)f$ is invertible since $f \in \mathrm{rad}\,(X, Y)$. Then $1_X - h(gf)$ is invertible, which shows that $gf \in \mathrm{rad}_{\mathscr{C}}$. This shows that $\mathrm{rad}_{\mathscr{C}}$ satisfies condition (3a) of Definition 7.4, and condition (3b) can be shown in a similar way. □

Lemma 7.6. $\mathrm{rad}_{\mathscr{C}}(X, X)$ *is the radical of the algebra* $\mathrm{End}\, X = \mathrm{Hom}_{\mathscr{C}}(X, X)$.

Proof. This follows directly from Lemma 4.1. □

Now let us consider the category $\mathrm{mod}\, A$. Our goal is to characterize irreducible morphisms in terms of the radical of $\mathrm{mod}\, A$. For simplicity, we will write rad and $\mathrm{rad}\,(X, Y)$ instead of $\mathrm{rad}_{\mathrm{mod}\, A}$ and $\mathrm{rad}_{\mathrm{mod}\, A}(X, Y)$.

Lemma 7.7. *If* X, Y *are indecomposable* A*-modules, then* $\mathrm{rad}\,(X, Y)$ *is equal to*

$$\{f \in \mathrm{Hom}(X, Y) \mid f \text{ is not a section and not a retraction.}\}$$

Proof. ($\subseteq$) If $f \in \mathrm{rad}\,(X, Y)$, then f is not an isomorphism. Since X and Y are indecomposable, f cannot be a section, neither a retraction.

($\supseteq$) Let $f \in \mathrm{Hom}(X, Y)$ such that f is not a section and not a retraction, and let $g \in \mathrm{Hom}(Y, X)$. Then $gf \in \mathrm{End}\, X$. Corollary 4.20 implies that the algebras $\mathrm{End}\, X$ and $\mathrm{End}\, Y$ are local, since X and Y are indecomposable. We will start by showing that gf is not an isomorphism.

Suppose the contrary, then there exists $h \in \mathrm{End}\, X$ such that $hgf = 1_X$.

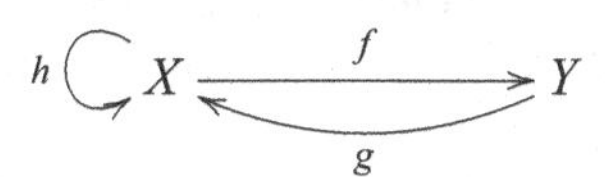

Then $1_X = (hgf)(hgf) = hg(fhg)f$ which shows that fhg is a nonzero element of $\mathrm{End}\, Y$ and $(fhg)^2 = f(hgf)hg = fhg$. Thus fhg is a nonzero idempotent in $\mathrm{End}\, Y$, which implies that $fhg = 1_Y$, a contradiction to the hypotheses that f is not a retraction.

Now, since gf is not an isomorphism, it follows that gf has no right inverse in the local algebra $\mathrm{End}\, X$ and therefore is contained in the unique maximal ideal $\mathrm{rad}\,\mathrm{End}\, X$ of $\mathrm{End}\, X$. By Lemma 4.1, it follows that $(1_X - gf)$ has a two-sided inverse, henceis an isomorphism. This shows that $f \in \mathrm{rad}\,(X, Y)$. □

Let $\mathrm{rad}^2(X, Y)$ be the span of all morphisms $f : X \to Y$ for which we have a factorization $f = gh$ with $g \in \mathrm{rad}\,(Z, Y)$ and $h \in \mathrm{rad}\,(X, Z)$ for some module Z. Clearly $\mathrm{rad}^2(X, Y) \subset \mathrm{rad}\,(X, Y)$.

With these notions at hand, we can characterize irreducible morphisms as follows:

Lemma 7.8. *Let X, Y be indecomposable A-modules. Then $f : X \to Y$ is irreducible if and only if $f \in \operatorname{rad}(X, Y) \setminus \operatorname{rad}^2(X, Y)$.*

Proof. ($\Rightarrow$) If f is irreducible, then $f \in \operatorname{rad}(X, Y)$, by Lemma 7.7. Suppose that $f \in \operatorname{rad}^2(X, Y)$. Then there is a commutative diagram

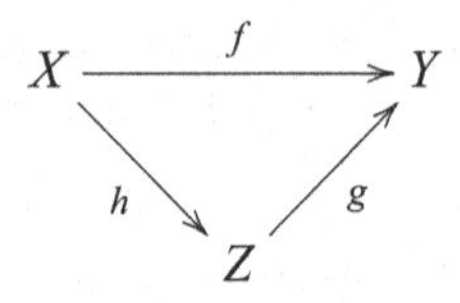

for some $g \in \operatorname{rad}(Z, Y)$ and $h \in \operatorname{rad}(X, Z)$, and Z is a sum $Z = \oplus_{i=1}^t Z_i$ with each Z_i indecomposable. Then h and g can be written in matrix form as

$$h = \begin{bmatrix} h_1 \\ h_2 \\ \vdots \\ h_t \end{bmatrix}, \quad g = \begin{bmatrix} g_1 \, g_2 \cdots g_t \end{bmatrix},$$

with $h_i : X \to Z_i$ and $g_i : Z_i \to Y$.

Since f is irreducible, it follows that h is a section or g is a retraction. If h is a section, let

$$h' = \begin{bmatrix} h_1' \, h_2' \cdots h_t' \end{bmatrix} : \oplus Z_i \to X$$

be such that $1_X = h'h$. Then, for all i, we have $h_i \in \operatorname{rad}(X, Z_i)$, and thus $h_i' h_i \in \operatorname{rad}(X, X) = \operatorname{rad}\operatorname{End}(X)$, and therefore $1_X = h'h = \sum_i h_i' h_i \in \operatorname{rad}\operatorname{End}(X)$, which contradicts the fact that $\operatorname{rad}\operatorname{End}(X) \neq \operatorname{End}(X)$. Thus h is not a section. In a similar way, one can prove that g is not a retraction, and therefore f cannot be in $\operatorname{rad}^2(X, Y)$.

($\Leftarrow$) Let $f \in \operatorname{rad}(X, Y) \setminus \operatorname{rad}^2(X, Y)$. Since X and Y are indecomposable and f is not an isomorphism, it follows that f is not a section and not a retraction either. Suppose that $f = gh$ with $h : X \to Z$ and $g : Z \to Y$. Say $Z = \oplus_{i=1}^t Z_i$ with each Z_i indecomposable. Then again, h and g can be written in matrix form as

$$h = \begin{bmatrix} h_1 \\ h_2 \\ \vdots \\ h_t \end{bmatrix}, \quad g = \begin{bmatrix} g_1 \, g_2 \cdots g_t \end{bmatrix},$$

with $h_i : X \to Z_i$ and $g_i : Z_i \to Y$, and we have

$$f = gh = \sum_{i=1}^{t} g_i h_i.$$

Since $f \notin \operatorname{rad}^2(X, Y)$, one of the h_i or one of the g_j must be invertible. But if h_i is invertible, then $1_X = \begin{bmatrix} 0 \cdots 0 \; h_i^{-1} \; 0 \cdots 0 \end{bmatrix} h$, which shows that h is a section, and if g_j is invertible, then

$$1_Y = g \begin{bmatrix} 0 \\ \vdots \\ 0 \\ g_j^{-1} \\ 0 \\ \vdots \\ 0 \end{bmatrix},$$

which shows that g is a retraction. This proves that f is irreducible. □

We are now able to give the definition of the Auslander–Reiten quiver.

Definition 7.6. Let A be a finite-dimensional algebra. The **Auslander–Reiten quiver** of A is the quiver whose vertices consist of isoclasses of indecomposable A-modules; and if X and Y are indecomposable A-modules, then the number of arrows from X to Y is equal to the dimension of $\operatorname{rad}(X, Y)/\operatorname{rad}^2(X, Y)$.

7.2 Auslander–Reiten Translation

We have already used the Auslander–Reiten translation τ in the context of quiver representations. Here we consider the more general setting of modules over a finite-dimensional k-algebra. We start with the definition of τ in this situation which is analogous to its definition for quiver representations.

Let $\operatorname{proj} A$ (respectively $\operatorname{inj} A$) denote the full subcategory of $\operatorname{mod} A$ whose objects are the projective modules (respectively the injective modules). According to Proposition 2.29 and Exercise 5.6 of Chap. 5, the Nakayama functor $\nu = D\operatorname{Hom}_A(-, A) : \operatorname{mod} A \to \operatorname{mod} A$ induces an equivalence of categories

$$\nu : \operatorname{proj} A \xrightarrow{\nu} \operatorname{inj} A,$$

whose quasi-inverse is given by $\nu^{-1} = \operatorname{Hom}_A(DA, -)$, where $DA = \oplus_{i \in Q_0} I(i)$ is the direct sum of the indecomposable injective modules. An exact sequence of A-modules

$$P_1 \xrightarrow{p_1} P_0 \xrightarrow{p_0} M \longrightarrow 0,$$

with P_0, P_1 projective, is called a **projective presentation** of M. Dually an exact sequence

$$0 \longrightarrow M \longrightarrow I_0 \longrightarrow I_1 ,$$

with I_0 and I_1 injective, is called an **injective presentation** of M. In other words, a projective presentation is the beginning of a projective resolution. We call the projective presentation *minimal* if the corresponding projective resolution is.

Let M be an A-module. The Auslander–Reiten translation τM of M is defined as follows: Start with a minimal projective presentation of M

$$P_1 \xrightarrow{p_1} P_0 \xrightarrow{p_0} M \longrightarrow 0.$$

Thus $P_0 \xrightarrow{p_0} M$ is a projective cover and $P_1 \xrightarrow{p_0} \ker p_0$ is a projective cover. Applying the Nakayama functor yields an exact sequence

$$0 \longrightarrow \tau M \longrightarrow \nu P_1 \xrightarrow{\nu p_1} \nu P_0 \xrightarrow{\nu p_0} \nu M \longrightarrow 0$$

where $\tau M = \ker \nu p_1$ is the **Auslander–Reiten translate** of M.

Dually, start with a minimal injective presentation

$$0 \longrightarrow M \xrightarrow{i_0} I_0 \xrightarrow{i_1} I_1$$

and apply ν^{-1} to get an exact sequence

$$0 \longrightarrow \nu^{-1}M \xrightarrow{\nu^{-1}i_0} \nu^{-1}I_0 \xrightarrow{\nu^{-1}i_1} \nu^{-1}I_1 \longrightarrow \tau^{-1}M \longrightarrow 0,$$

where $\tau^{-1}M = \operatorname{coker} \nu^{-1}i_1$ is the **inverse Auslander–Reiten translate of** M.

Remark 7.9. If M is projective, then $\tau M = 0$, since in the minimal projective presentation above $M = P_0$ and $P_1 = 0$.

Dually, if M is injective, then $\tau^{-1}M = 0$, since in the minimal injective presentation above $M = I_0$ and $I_1 = 0$.

The following lemma gives a very useful characterization of modules of projective dimension at most 1 in terms of their Auslander–Reiten translates:

Lemma 7.10. *Let M be an A-module. Then*

1. $\operatorname{pd} M \leq 1 \Leftrightarrow \operatorname{Hom}(DA, \tau M) = 0$,
2. $\operatorname{id} M \leq 1 \Leftrightarrow \operatorname{Hom}(\tau^{-1}M, A) = 0$.

Proof. Since the two statements are dual, we will only prove (1). Let

$$P_1 \xrightarrow{p_1} P_0 \xrightarrow{p_0} M \longrightarrow 0.$$

be a minimal projective presentation. Applying the Nakayama functor yields the exact sequence

$$0 \longrightarrow \tau M \longrightarrow \nu P_1 \xrightarrow{\nu p_1} \nu P_0 \xrightarrow{\nu p_0} \nu M \longrightarrow 0$$

Now applying the left exact functor $\nu^{-1} = \mathrm{Hom}(DA, -)$ and using the isomorphism $\nu^{-1}\nu|_{\mathrm{proj}\,A} \cong 1_{\mathrm{proj}\,A}$ gives us the following commutative diagram with exact rows:

$$\begin{array}{ccccccccccc}
0 & \longrightarrow & \mathrm{Hom}_A(DA, \tau M) & \longrightarrow & \nu^{-1}\nu P_1 & \xrightarrow{\nu^{-1}\nu p_1} & \nu^{-1}\nu P_0 & & & & \\
\downarrow\cong & & & & \downarrow\cong & & \downarrow\cong & & & & \\
0 & \longrightarrow & \ker p_1 & \longrightarrow & P_1 & \xrightarrow{p_1} & P_0 & \longrightarrow & M & \longrightarrow & 0
\end{array}$$

The universal property of the kernel implies that there exists a morphism

$$h\colon \mathrm{Hom}_A(DA, \tau M) \cong \ker p_1,$$

and it follows from the Five Lemma (Lemma 4.11) that h is an isomorphism. Therefore $\mathrm{Hom}_A(DA, \tau M) = 0$ if and only if p_1 is injective if and only if $\mathrm{pd}\, M \leq 1$. □

Example 7.1. Consider the algebra given by the quiver

$$1 \xrightarrow{\alpha} 2 \xrightarrow{\beta} 3 \xrightarrow{\gamma} 4 \xrightarrow{\delta} 5$$

bound by $\alpha\beta = 0, \gamma\delta = 0$. Its Auslander–Reiten quiver is given by

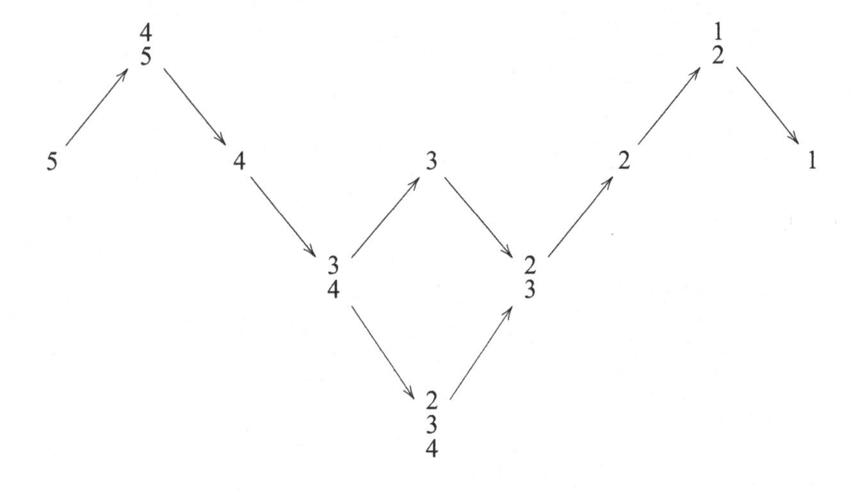

and $DA = \begin{smallmatrix}4\\5\end{smallmatrix} \oplus \begin{smallmatrix}2\\3\\4\end{smallmatrix} \oplus \begin{smallmatrix}2\\3\end{smallmatrix} \oplus \begin{smallmatrix}1\\2\end{smallmatrix} \oplus 1$. From the Auslander–Reiten quiver it is clear that for any indecomposable A-module N, $\mathrm{Hom}(DA, N) \neq 0$ if and only if N is either one of the five injective direct summands of DA or $N = 4$ or $N = \begin{smallmatrix}3\\4\end{smallmatrix}$ or $N = 2$. Since τM cannot be injective, we see that $\mathrm{Hom}(DA, \tau M) \neq 0$ if and only if $M = 3$ or $M = \begin{smallmatrix}2\\3\end{smallmatrix}$ or $M = 1$. Therefore the lemma implies that the three modules 3, $\begin{smallmatrix}2\\3\end{smallmatrix}$, and 1 are the only modules of projective dimension greater than 1.

7.3 Coxeter Transformation

We have already used the Coxeter transformation for computing τ-orbits in the Auslander–Reiten quivers of type $\mathbb{A}$ and $\mathbb{D}$ in Sects. 3.1.2.2 and 3.3.2.2. We are now ready to study the Coxeter transformation in a much more general setting.

Let A be a finite-dimensional algebra of finite global dimension, and let $e_1, \ldots, e_n$ be primitive orthogonal idempotents such that $1_A = e_1 + e_2 + \cdots + e_n$. The **Cartan matrix** of A is defined to be the $n \times n$ matrix:

$$C_A = \begin{bmatrix} b_{11} & \cdots & b_{1n} \\ \vdots & & \vdots \\ b_{n1} & \cdots & b_{nn} \end{bmatrix} \qquad \text{where } b_{ij} = \dim e_j A e_i.$$

Remark 7.11. If A is a bound quiver algebra $A = kQ/I$, then the ith column of C_A is the dimension vector of the indecomposable projective $P(i)$ at vertex i and the ith row of C_A is the dimension vector of the indecomposable injective $I(i)$ at vertex i.

Example 7.2. Let Q be the quiver

$$1 \xrightarrow{\ \alpha\ } 2 \xrightarrow{\ \beta\ } 3 \xrightarrow{\ \gamma\ } 4$$

bound by the relation $\alpha\beta\gamma = 0$. Then

$$C_A = \begin{bmatrix} 1&0&0&0 \\ 1&1&0&0 \\ 1&1&1&0 \\ 0&1&1&1 \end{bmatrix}.$$

Note that C_A has determinant 1; thus we can compute its inverse:

$$C_A^{-1} = \begin{bmatrix} 1 & 0 & 0 & 0 \\ -1 & 1 & 0 & 0 \\ 0 & -1 & 1 & 0 \\ 1 & 0 & -1 & 1 \end{bmatrix}.$$

Proposition 7.12. *Let $A = kQ/I$ be a bound quiver algebra of finite global dimension. Then*

$$\det C_A \in \{1, -1\}.$$

In particular, C_A is invertible.

Proof. Let i be an arbitrary vertex of Q and $S(i)$ the corresponding simple module. Since gldim $A < \infty$, there exists finite projective resolution, say

$$0 \longrightarrow P_{m_i} \longrightarrow \cdots \longrightarrow P_1 \longrightarrow P_0 \longrightarrow S(i) \longrightarrow 0.$$

Then, since this is an exact sequence, we have

$$\underline{\dim}\, S(i) = \sum_{j=0}^{m_i} (-1)^j \underline{\dim}\, P_j.$$

Since the projectives P_j on the right-hand side of this equation are direct sums of indecomposable projectives, then it follows by Remark 7.11 that there exists a vector $d(i) \in \mathbb{Z}^n$ such that the right-hand side is equal to $C_A d(i)$. On the other hand, $\underline{\dim}\, S(i) = (0, \dots, 0, 1, 0, \dots, 0)$ with the 1 in the ith position. Now if we let the vertex i vary, we can write the identity $n \times n$ matrix as

$$1 = [C_A(d(i))]_{i=1,\dots,n} = C_A\, D,$$

where the matrix D has columns $d(1), \dots, d(n)$. This shows that C_A is invertible. Moreover, since the entries of both matrices C_A and D are integers, their determinants are two integers whose product is equal to 1; thus $\det C_A$ is either 1 or -1. □

Knowing that C_A is invertible, we can use its inverse to define another matrix.

Definition 7.7. The **Coxeter matrix** Φ_A of A is the $n \times n$ integer matrix:

$$\Phi_A = -C_A^t C_A^{-1}.$$

The **Coxeter transformation** is the linear map $\mathbb{Z}^n \to \mathbb{Z}^n$ defined by $x \mapsto \Phi_A(x)$.

Example 7.3. In Example 7.2, the Coxeter matrix is

$$-\begin{bmatrix} 1&1&1&0\\ 0&1&1&1\\ 0&0&1&1\\ 0&0&0&1 \end{bmatrix}\begin{bmatrix} 1&0&0&0\\ -1&1&0&0\\ 0&-1&1&0\\ 1&0&-1&1 \end{bmatrix}=\begin{bmatrix} 0&0&-1&0\\ 0&0&0&-1\\ -1&1&0&-1\\ -1&0&1&-1 \end{bmatrix}.$$

Lemma 7.13. $\Phi_A(\underline{\dim}\, P(i)) = -\underline{\dim}\, I(i)$.

Proof. By Remark 7.11, we have $\underline{\dim}\, P(i) = C_A(\underline{\dim}\, S(i))$, which implies that

$$-\Phi_A(\underline{\dim}\, P(i)) = C_A^t C_A^{-1} C_A(\underline{\dim}\, S(i)) = C_A^t(\underline{\dim}\, S(i)),$$

and this last vector is equal to the ith row of C_A, which, again by Remark 7.11, is equal to $\underline{\dim}\, I(i)$. □

Remark 7.14. If A is a path algebra, thus Q has no oriented cycles, and the ideal I is trivial, then the Coxeter transformation is given as a sequence of simple reflections, the *Coxeter element*, $\Phi_A = s_{i_1} s_{i_2} \cdots s_{i_n}$ where $(i_1, \ldots, i_n)$ is an admissible sequence; compare with Sects. 3.1 and 3.3.

Now we use the Coxeter transformation to compute the dimension vector of the Auslander–Reiten translate of a given module.

Proposition 7.15. *Let A be a finite-dimensional algebra of finite global dimension.*

1. *Let M be an indecomposable, non-projective A-module and let*

$$P_1 \xrightarrow{p_1} P_0 \xrightarrow{p_0} M \longrightarrow 0$$

be a minimal projective presentation. Then

$$\underline{\dim}\tau\, M = \Phi_A\, \underline{\dim}\, M - \Phi_A\, \underline{\dim}\, \ker p_1 + \underline{\dim}\, \nu M.$$

2. *Let M be an indecomposable, non-injective A-module and let*

$$0 \longrightarrow M \xrightarrow{i_0} I_0 \xrightarrow{i_1} I_1$$

be a minimal injective presentation. Then

$$\underline{\dim}\, \tau^{-1} M = \Phi_A^{-1}\, \underline{\dim}\, M - \Phi_A^{-1}\, \underline{\dim}\, \mathrm{coker}\, i_1 + \underline{\dim}\, \nu^{-1} M.$$

Proof. We only prove (1), since (2) is the dual statement and its proof is similar. Let us recall first that the Nakayama functor ν maps the indecomposable projective $P(i)$

to the corresponding indecomposable injective $I(i)$; see Sect. 2.3 or Exercise 5.6 in Chap. 5. Thus, by Lemma 7.13, we have $\Phi_A \underline{\dim}\, P = -\underline{\dim}\, \nu P$ for all projective modules P.

Now, the exact sequence

$$0 \longrightarrow \ker p_1 \longrightarrow P_1 \xrightarrow{p_1} P_0 \xrightarrow{p_0} M \longrightarrow 0$$

yields

$$\underline{\dim}\, M - \underline{\dim}\, \ker p_1 = \underline{\dim}\, P_0 - \underline{\dim}\, P_1,$$

and so, by applying Φ_A and using the remark above, we get

$$\Phi_A \underline{\dim}\, M - \Phi_A \underline{\dim}\, \ker p_1 = -\underline{\dim}\, \nu P_0 + \underline{\dim}\, \nu P_1. \tag{7.1}$$

On the other hand, the exact sequence

$$0 \longrightarrow \tau M \longrightarrow \nu P_1 \longrightarrow \nu P_0 \longrightarrow \nu M \longrightarrow 0$$

yields

$$\underline{\dim}\, \tau M = \underline{\dim}\, \nu P_1 - \underline{\dim}\, \nu P_0 + \underline{\dim}\, \nu M,$$

and now the statement follows from equation (7.1). □

Example 7.4. Let us continue the Examples 7.2 and 7.3. A is the bound quiver algebra $1 \xrightarrow{\alpha} 2 \xrightarrow{\beta} 3 \xrightarrow{\gamma}$ bound by the relation $\alpha\beta\gamma = 0$, and its Coxeter matrix is

$$\Phi_A = \begin{bmatrix} 0 & 0 & -1 & 0 \\ 0 & 0 & 0 & -1 \\ -1 & 1 & 0 & -1 \\ -1 & 0 & 1 & -1 \end{bmatrix}.$$

Let $M = I(2) = \begin{smallmatrix} 1 \\ 2 \end{smallmatrix}$, and suppose we want to compute τM. Start with a projective presentation:

$$\begin{smallmatrix} 3 \\ 4 \end{smallmatrix} \xrightarrow{p_1} \begin{smallmatrix} 1 \\ 2 \\ 3 \end{smallmatrix} \longrightarrow \begin{smallmatrix} 1 \\ 2 \end{smallmatrix} \longrightarrow 0\,.$$

By direct inspection, we can see that $\operatorname{Hom}(M, A) = 0$; thus $\nu M = 0$. Therefore, Proposition 7.15 implies that

$$\underline{\dim}\, \tau M = \Phi_A \begin{bmatrix} 1 \\ 1 \\ 0 \\ 0 \end{bmatrix} - \Phi_A \begin{bmatrix} 0 \\ 0 \\ 0 \\ 1 \end{bmatrix} + 0 = \begin{bmatrix} 0 \\ 0 \\ 0 \\ -1 \end{bmatrix} - \begin{bmatrix} 0 \\ -1 \\ -1 \\ -1 \end{bmatrix} = \begin{bmatrix} 0 \\ 1 \\ 1 \\ 0 \end{bmatrix},$$

which shows that $\tau \begin{smallmatrix} 1 \\ 2 \end{smallmatrix} = \begin{smallmatrix} 2 \\ 3 \end{smallmatrix}$.

This agrees with the results obtained in Example 5.12.

Corollary 7.16. *Let A be a finite-dimensional algebra of finite global dimension.*

1. *Let M be an indecomposable A-module such that* $\operatorname{pd} M \leq 1$ *and* $\operatorname{Hom}(M, A) = 0$*; then*

$$\underline{\dim}\, \tau M = \Phi_A \, \underline{\dim}\, M.$$

2. *Let M be an indecomposable A-module such that* $\operatorname{id} M \leq 1$ *and* $\operatorname{Hom}(DA, M) = 0$*; then*

$$\underline{\dim}\, \tau^{-1} M = \Phi_A^{-1} \, \underline{\dim}\, M.$$

This corollary applies in particular to non-projective (respectively non-injective) indecomposable modules M over hereditary algebras, since then the conditions $\operatorname{pd} M \leq 1$ and $\operatorname{Hom}(M, A) = 0$ (respectively $\operatorname{id} M \leq 1$ and $\operatorname{Hom}(DA, M) = 0$) are always satisfied. We have used this fact in order to construct Auslander–Reiten quivers in Chap. 3.

Proof. We only show (1). The conditions in the corollary imply that $\ker p_1 = 0$ and $\nu M = 0$, and then the result follows immediately from Lemma 7.15. □

7.4 Auslander–Reiten Formulas

In this section, we will prove the Auslander–Reiten formulas. This is a fundamental result which, on the one hand, is of conceptual nature, since it describes a relation between short exact sequences and morphisms in the module category, and, on the other hand, also provides a powerful computational tool, since it provides a way to calculate Ext^1 in terms of morphisms.

We start with a result about the Auslander–Reiten translation τ.

Proposition 7.17. *Let M and N be indecomposable A-modules:*

1. *$\tau M = 0$ if and only if M is projective.*
2. *If M is non-projective, then τM is indecomposable and $\tau^{-1}\tau M = M$.*
3. *If M and N are non-projective, then*

$$M \cong N \iff \tau N \cong \tau M.$$

Proof. Let $P_1 \xrightarrow{p_1} P_0 \longrightarrow M \longrightarrow 0$ be a minimal projective presentation.

(1) If M is projective, then $P_1 = 0$ which implies that $\tau\, M = 0$. On the other hand, if $\tau\, M = 0$, then $\nu p_1 : \nu P_1 \to \nu P_0$ is injective. Consider the diagram

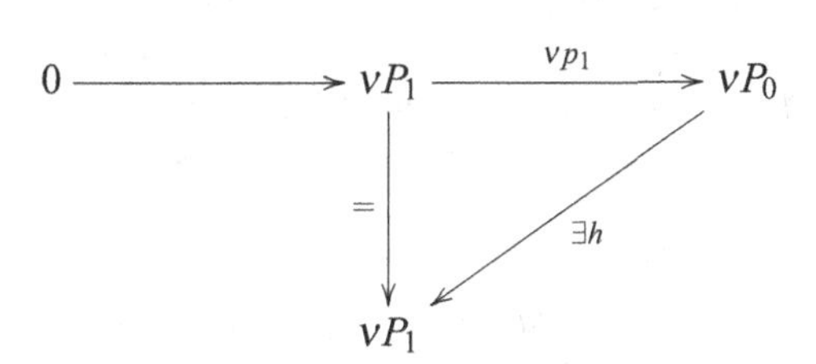

where the map h exists because νP_1 is injective. It follows that νp_1 is a section and hence that νP_1 is a direct summand of νP_0; thus P_1 is a direct summand of P_0. But since the projective presentation is minimal, this implies that $P_1 = 0$, and thus M is projective.

(2) Let us show first that p_1 is an indecomposable map, that is, there is no decomposition $P_1 = L_1 \oplus N_1$ and $P_2 = L_2 \oplus N_2$ such that the morphism p_1 decomposes diagonally as

$$p_1 = \begin{bmatrix} f & 0 \\ 0 & g \end{bmatrix},$$

with $f = p_1|_{L_1}$ and $g = p_1|_{N_1}$. Indeed, suppose such a decomposition exists, then $M = \operatorname{coker} p_1 = \operatorname{coker} f \oplus \operatorname{coker} g$ and, since M is indecomposable, we must have that either f or g is surjective, but this contradicts the minimality of the projective presentation.

Now suppose that $\tau M = X \oplus Y$. Since

$$0 \longrightarrow \tau M \longrightarrow \nu P_1 \xrightarrow{\nu p_1} \nu P_0$$

is a minimal injective presentation, it follows that νp_1 is a decomposable map, but since $\nu : \operatorname{proj} A \to \operatorname{inj} A$ is an equivalence of categories, it follows that p_1 too is decomposable, a contradiction. This shows that τM is indecomposable.

In order to show that $\tau^{-1}\tau M = M$, we apply ν^{-1} to the minimal injective presentation

$$0 \longrightarrow \tau M \longrightarrow \nu P_1 \longrightarrow \nu P_0$$

and get an exact sequence

$$\nu^{-1}\tau M \longrightarrow \nu^{-1}\nu P_1 \xrightarrow{\nu^{-1}\nu p_1} \nu^{-1}\nu P_0 \longrightarrow \tau^{-1}\tau M \longrightarrow 0.$$

Since $\nu : \operatorname{proj} A \to \operatorname{inj} A$ is an equivalence of categories, it follows that $\tau^{-1}\tau M = \operatorname{coker} p_1 = M$.

(3) If $M \cong N$ then M and N have isomorphic minimal projective presentations and thus $\tau M \cong \tau N$. On the other hand, if $\tau M \cong \tau N$, then τM and τN have isomorphic minimal injective presentations and thus $M = \tau^{-1}\tau M \cong \tau^{-1}\tau N = N$.

□

Categories 8 *In order to state and prove the Auslander–Reiten formulas, we need the following definitions:*

Definition 7.8. *Let $F, G: \mathscr{C} \to \mathscr{D}$ be two covariant functors between two categories. A **functorial morphism** $\omega: F \to G$ is a family of morphisms $\omega_L : F(L) \to G(L)$ where L runs over all objects in the category $\mathscr{C}$, such that, for any morphism $f: L_1 \to L_2$ in $\mathscr{C}$, the following diagram is commutative:*

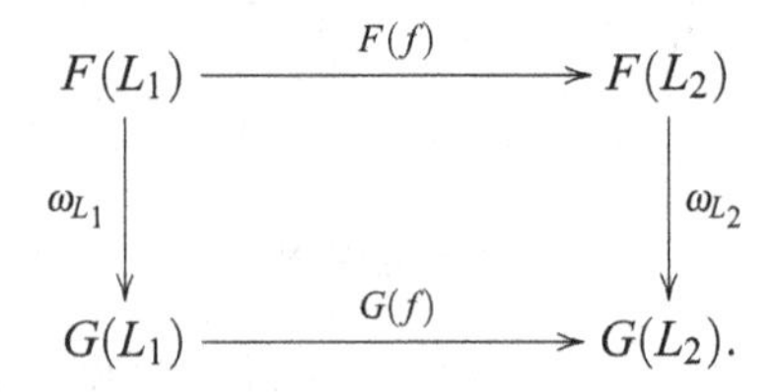

*Dually, if $F, G: \mathscr{C} \to \mathscr{D}$ are two contravariant functors, then a **functorial morphism** $\omega: F \to G$ is a family of morphisms $\omega^L : F(L) \to G(L)$ where L runs over all objects in the category $\mathscr{C}$, such that, for any morphism $f: L_1 \to L_2$ in $\mathscr{C}$, the following diagram is commutative:*

$$\begin{array}{ccc} F(L_2) & \xrightarrow{F(f)} & F(L_1) \\ \downarrow{\scriptstyle \omega^{L_2}} & & \downarrow{\scriptstyle \omega^{L_1}} \\ G(L_2) & \xrightarrow{G(f)} & G(L_1). \end{array}$$

Definition 7.9. Let $P(X, Y)$ be the set of all morphisms $f \in \operatorname{Hom}(X, Y)$ such that f factors through a projective A-module, and define

$$\underline{\operatorname{Hom}}(X, Y) = \operatorname{Hom}(X, Y)/P(X, Y).$$

Dually, let $I(X, Y)$ be the set of all morphisms $f \in \operatorname{Hom}(X, Y)$ such that f factors through an injective A-module, and define

$$\overline{\operatorname{Hom}}(X, Y) = \operatorname{Hom}(X, Y)/I(X, Y).$$

We are ready for the main result of this section.

Theorem 7.18 (Auslander–Reiten formulas). *Let M, N be A-modules. Then there are isomorphisms*

$$\operatorname{Ext}^1(M, N) \cong D\underline{\operatorname{Hom}}(\tau^{-1}N, M) \cong D\overline{\operatorname{Hom}}(N, \tau M)$$

that are functorial in both variables, that is, the following functors are isomorphic:

$$\begin{aligned}\operatorname{Ext}^1(-, N) &\cong D\underline{\operatorname{Hom}}(\tau^{-1}N, -) \cong D\overline{\operatorname{Hom}}(N, \tau -)\\ \operatorname{Ext}^1(M, -) &\cong D\underline{\operatorname{Hom}}(\tau^{-1}-, M) \cong D\overline{\operatorname{Hom}}(-, \tau M).\end{aligned}$$

The proof of the theorem is rather involved and requires two lemmas which use the concept of the tensor product of two modules. Since we are not using the tensor product elsewhere in the book, we refrain from giving a rigorous introduction to tensor products and instead only state the definition and the properties that we need in the proof.

7.4.1 Tensor Products

In this subsection, we give a brief overview of tensor products. For further details on tensor products, we refer the reader to Sect. 10.4 of [31].

Let A be a ring with 1, M be right A-module, N a left A-module, and G an abelian group.

Definition 7.10. A map $h : M \times N \to G$ is called *A-balanced* if for all $m, m' \in M$, $n, n' \in N$ and $a \in A$, we have

$$\begin{aligned}h(m + m', n) &= h(m, n) + h(m', n);\\ h(m, n + n') &= h(m, n) + h(m, n');\\ h(m, an) &= h(ma, n).\end{aligned}$$

Definition 7.11. The *tensor product* of M and N over A is an abelian group $M \otimes_A N$ together with an A-balanced map $i : M \times N \to M \otimes_A N$, such that, for every abelian group G and every A-balanced map $h : M \times N \to G$, there exists a unique group homomorphism $h' : M \otimes_A N \to G$ such that $h' \circ i = h$.

One can show the existence of the tensor product by giving an explicit construction as a quotient of the free module with basis $M \times N$. One can also show that the tensor product of two modules is unique up to isomorphism.

The elements $i(m, n)$ of the tensor product $M \otimes_A N$ are called simple tensors and are denoted as $m \otimes n = i(m, n)$. Arbitrary elements of $M \otimes_A N$ are finite sums of simple tensors.

If M is also an A-bimodule, then $M \otimes N$ is a left A-module with respect to the A-action:

$$a(m \otimes n) = (am) \otimes n.$$

Similarly, if N is an A-bimodule, then $M \otimes N$ is a right A-module with respect to the A-action:

$$(m \otimes n)a = m \otimes (na).$$

The tensor product has the following properties:

Proposition 7.19. *Let M, M' be right A-modules and N, N' left A-modules.*

1. (identity)

$$A \otimes_A N \cong N \quad and \quad M \otimes_A A \cong M.$$

2. (associativity)

$$M \otimes_A (N \oplus N') \cong (M \otimes_A N) \oplus (M \otimes_A N')$$

and

$$(M \oplus M') \otimes_A N \cong (M \otimes_A N) \oplus (M' \otimes_A N).$$

Proposition 7.20. *Let A, B be rings with 1, L a right A-module, M an (A, B)-bimodule, and N a right B-module. Then there is an isomorphism of groups:*

$$\mathrm{Hom}_B(L \otimes_A M, N) \cong \mathrm{Hom}_A(L, \mathrm{Hom}_B(M, N)). \tag{7.2}$$

The isomorphism in Proposition 7.20 is called *adjoint isomorphism*.

We proceed now with the proof of the Auslander–Reiten formulas, Theorem 7.18. We will need the following two lemmas:

Lemma 7.21. *Let L, M be A-modules and let*

$$\varphi_M^L : M \otimes_A \mathrm{Hom}_A(L, A) \to \mathrm{Hom}_A(L, M)$$

be the group homomorphism defined on simple tensors by $\varphi_M^L(m \otimes f) = (\ell \mapsto mf(\ell))$. Then

1. *If M is projective, then φ_M^L is an isomorphism.*
2. $\mathrm{coker}\, \varphi_M^L = \underline{\mathrm{Hom}}_A(L, M)$.
3. *φ_M^- is a functorial morphism*

$$\varphi_M^- : M \otimes_A \mathrm{Hom}_A(-, A) \to \mathrm{Hom}_A(-, M).$$

Proof.

(1) If M is projective indecomposable, and thus $M = eA$ for some primitive idempotent $e \in A$, then

$$\begin{aligned} M \otimes_A \mathrm{Hom}_A(L, A) &= e(A \otimes_A \mathrm{Hom}_A(L, A) \\ &\cong e\mathrm{Hom}_A(L, A) \\ &\cong \mathrm{Hom}_A(L, eA) \\ &\cong \mathrm{Hom}_A(L, M), \end{aligned}$$

where we used Proposition 7.19 and the left A-module structure of $\mathrm{Hom}_A(L, A)$. Moreover, under this sequence of isomorphism, the element $ea \otimes f$ is mapped to the morphism $\ell \mapsto eaf(\ell)$, which shows the result for indecomposable projectives. If M is projective but not indecomposable, then the result follows from the additivity of Hom and the distributivity of the tensor product.

(2) We want to show that $\mathrm{im}\, \varphi_M^L = P(L, M)$.

($\subseteq$) Let h denote a map $\ell \mapsto mf(\ell)$ in the image of φ_M^L, and let $p : P \to M$ be a surjective morphism with P projective. Let $x \in P$ be such that $p(x) = m$, and define $g : L \to P$ by $g(\ell) = xf(\ell)$, where the right-hand side is the right action of the algebra element $f(\ell) \in A$ on the module element $x \in P$. Then

$$pg(\ell) = p(xf(\ell)) = p(x)f(\ell) = mf(\ell) = h(\ell).$$

Thus h factors through the projective module P; whence $h \in P(L, M)$.

($\supseteq$) Let $h \in P(L, M)$. Then there is a commutative diagram

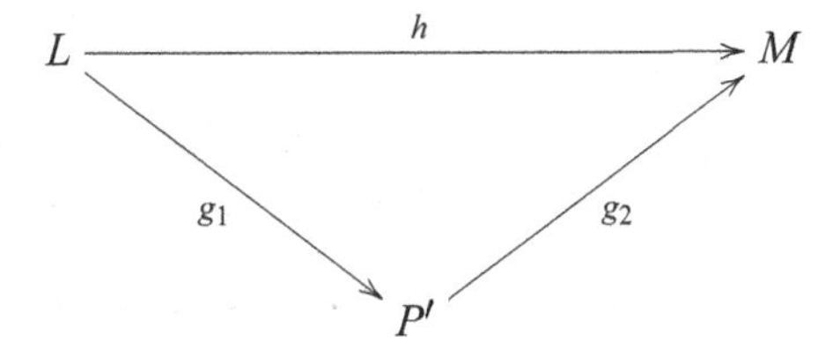

with P' projective. Let $p : P \to M$ be as above. Since p is surjective, the fact that P' is projective implies that there exists a morphism $u : P' \to P$ such that $g_2 = pu$. It follows that $h = pug_1$ with $(ug_1) \in \mathrm{Hom}(L, P)$. By part (1), there exists $x \in P$ and $f \in \mathrm{Hom}_A(L, A)$ such that $ug_1 = \varphi_P^L(x \otimes f)$, and therefore

$$h(\ell) = p(ug_1)(\ell) = p(xf(\ell)) = p(x)f(\ell),$$

with $p(x) \in M$. It follows that $h = \varphi_M^L(p(x) \otimes f)$; thus $h \in \mathrm{im}\,\varphi_M^L$. This proves (2), and (3) is an exercise.

□

Lemma 7.22. *There exists a morphism*

$$\omega_M^L : D\mathrm{Hom}_A(L, M) \to \mathrm{Hom}_A(M, \nu L)$$

such that

1. ω_M^P *is an isomorphism if* P *is projective.*
2. $\ker \omega_M^L = D\underline{\mathrm{Hom}}_A(L, M)$.
3. ω_M^- *is a functorial morphism*

$$\omega_M^-: D\mathrm{Hom}_A(-, M) \to \mathrm{Hom}_A(M, \nu -).$$

Proof. Now let $D\varphi_M^L : D\mathrm{Hom}_A(L, M) \to D(M \otimes_A \mathrm{Hom}_A(L, A))$ be the dual of the morphism in Lemma 7.21. Note that, by definition of the duality D, the target space of this morphism is equal to

$$\mathrm{Hom}_k(M \otimes_A \mathrm{Hom}_A(L, A), k),$$

and this is isomorphic to $\mathrm{Hom}_A(M, \mathrm{Hom}_k(\mathrm{Hom}_A(L, A), k)$, by equation (7.2), and this in turn is equal to $\mathrm{Hom}_A(M, D\mathrm{Hom}_A(L, A)) = \mathrm{Hom}_A(M, \nu L)$. Define ω_M^L to be the composition of $D\varphi_M^L$ with this isomorphism. Then (1) and (2) follow from Lemma 7.21 (1) and (2), respectively. Again, we leave the proof of statement (3) as an exercise. □

Proof of Theorem 7.18. We will only show that $\mathrm{Ext}^1(M, N) \cong D\underline{\mathrm{Hom}}(\tau^{-1}N, M)$. Without loss of generality, we may assume that N has no injective summands. Then $N = \tau L$ for some A-module L. Take a minimal projective presentation:

$$P_1 \xrightarrow{p_1} P_0 \xrightarrow{p_0} L \longrightarrow 0. \tag{7.3}$$

We apply the two functors $D\mathrm{Hom}(-, M)$ and $\mathrm{Hom}(M, \nu-)$ to the presentation (7.3), and then use the functorial morphism

$$\omega_M^-: D\mathrm{Hom}(-, M) \to \mathrm{Hom}(M, \nu -)$$

of Lemma 7.22 to compare the results. We then get a commutative diagram:

$$\begin{array}{ccccccc}
D\mathrm{Hom}(P_1,M) & \xrightarrow{\tilde{p}_1} & D\mathrm{Hom}(P_0,M) & \xrightarrow{\tilde{p}_0} & D\mathrm{Hom}(L,M) & \longrightarrow & 0\,, \\
\cong \downarrow \omega_M^{P_1} & & \cong \downarrow \omega_M^{P_0} & & \downarrow \omega_M^{L} & & \\
\mathrm{Hom}(M,\nu P_1) & \xrightarrow{\overline{p}_1} & \mathrm{Hom}(M,\nu P_0) & \xrightarrow{\overline{p}_0} & \mathrm{Hom}(M,\nu L) & &
\end{array}$$

where the top row is exact, since $D\mathrm{Hom}(-, M)$ is a right exact functor, the vertical morphisms $\omega_M^{P_1}$ and $\omega_M^{P_0}$ are isomorphisms by Lemma 7.22, and the bottom row is a complex, because, by construction, it is obtained by applying $\mathrm{Hom}(M, -)$ to the exact sequence

$$0 \longrightarrow N = \tau L \longrightarrow \nu P_1 \xrightarrow{\nu p_1} \nu P_0 \xrightarrow{\nu p_0} \nu L \to 0. \tag{7.4}$$

We want to compute $\mathrm{Ext}^1(M, N)$, knowing that $N = \tau L$. To do so, by definition of Ext^1, we apply the functor $\mathrm{Hom}(M, -)$ to the exact sequence (7.4), thereby obtaining the bottom row of our commutative diagram, and then

$$\mathrm{Ext}^1(M, N) = \ker \overline{p}_0 / \mathrm{im}\, \overline{p}_1. \tag{7.5}$$

In order to compute this quotient, we define a morphism $\psi\colon \ker \overline{p}_0 \to D\mathrm{Hom}(L, M)$ using the morphisms in the commutative diagram as follows:

$$\psi = \tilde{p}_0 \circ (\omega_M^{P_0})^{-1}|_{\ker \overline{p}_0} : \ker \overline{p}_0 \to D\mathrm{Hom}(L, M).$$

Thus ψ is the composition of the vertical map in the middle of the commutative diagram, but going upward, with the horizontal map $\tilde{p}_0$.

Claim 1: $\ker \psi = \mathrm{im}\, \overline{p}_1$.

To show claim 1, let $\overline{x} \in \ker \overline{p}_0$ such that $\psi(\overline{x}) = 0$. We then have $\left(\omega_M^{P_0}\right)^{-1}(\overline{x}) \in \ker \tilde{p}_0 = \mathrm{im}\, \tilde{p}_1$, and therefore $\overline{x} = \omega_M^{P_0} \tilde{p}_1(\tilde{y})$ for some $\tilde{y} \in D\mathrm{Hom}(P_1, M)$. Then the commutativity of our diagram implies that $\overline{x} = \overline{p}_1 \omega_M^{P_1}(\tilde{y}) \in \mathrm{im}\, \overline{p}_1$. This shows that $\ker \psi \subset \mathrm{im}\, \overline{p}_1$.

The other inclusion holds, because, if we have $\overline{x} = \overline{p}_1(\overline{y})$, then, using the commutativity of our diagram, we have $\psi(\overline{x}) = \tilde{p}_0 \tilde{p}_1 (\omega_M^{P_1})^{-1}(\overline{y})$, which is zero, since $\tilde{p}_0 \tilde{p}_1$ is zero. This proves claim 1.

Claim 2: $\mathrm{im}\, \psi = \ker \omega_M^L$.

If $\overline{x} \in \ker \overline{p}_0$, then $\omega_M^L(\psi(\overline{x})) = \overline{p}_0(\overline{x}) = 0$, which shows that $\mathrm{im}\, \psi \subset \ker \omega_M^L$.
To show the other inclusion, let $\tilde{u} \in \ker \omega_M^L$. Since $\tilde{p}_0$ is surjective, there exists $\tilde{x}$ such that $\tilde{u} = \tilde{p}_0(\tilde{x}) = \psi(\omega_M^{P_0}(\tilde{x})) \in \mathrm{im}\, \psi$. This proves claim 2.

Now claim 1 and 2 together with the first isomorphism theorem show that

$$\ker \omega_M^L \cong \ker \overline{p}_0 / \operatorname{im} \overline{p}_1.$$

By Lemma 7.22, the left-hand side of this equation is isomorphic to $D\underline{\mathrm{Hom}}(L, M)$ and, by equation (7.5), the right-hand side is $\mathrm{Ext}^1(M, N)$. □

The Auslander–Reiten formulas have the following very useful corollaries:

Corollary 7.23. *Let M, N be A-modules. Then*

1. *If* $\mathrm{pd}\, M \leq 1$, *then* $\mathrm{Ext}^1(M, N) \cong D\mathrm{Hom}(N, \tau M)$.
2. *If* $\mathrm{id}\, N \leq 1$, *then* $\mathrm{Ext}^1(M, N) \cong D\mathrm{Hom}(\tau^{-1} N, M)$.

Proof. Again, we only prove (1). Using Lemma 7.10, we see that $\mathrm{pd}\, M \leq 1$ implies that $\mathrm{Hom}(DA, \tau M) = 0$, which shows that $\overline{\mathrm{Hom}}(N, \tau M) = \mathrm{Hom}(N, \tau M)$. Now the result follows from Theorem 7.18. □

Corollary 7.24. *Let M and N be non-projective indecomposable A-modules such that* $\mathrm{pd}\, M \leq 1$ *and* $\mathrm{id}\, \tau N \leq 1$. *Then*

$$\mathrm{Hom}(\tau N, \tau M) \cong \mathrm{Hom}(N, M).$$

Proof. Since M is not projective, τM is an indecomposable A-module. By Corollary 7.23, we can compute $\mathrm{Ext}^1(M, \tau N)$ in two ways yielding $D\mathrm{Hom}(\tau N, \tau M) \cong \mathrm{Ext}^1(M, \tau N) \cong D\mathrm{Hom}(N, M)$. The result follows by applying the duality. □

Remark 7.25. Note that the conditions on the projective and injective dimension for the two corollaries are always satisfied if the algebra A is hereditary, because every module over a hereditary algebra has projective and injective dimension at most 1. We have applied the corollaries several times in Chap. 3 in order to compute dimensions of Ext in the Auslander–Reiten quivers of type $\mathbb{A}$ and $\mathbb{D}$.

Example 7.5. Let Q be the quiver

and let A be its path algebra. Thus A is hereditary and we can use Corollaries 7.23 and 7.24. Let M, N be the indecomposable A-modules:

$$M = \begin{smallmatrix} 1 \\ 4 \end{smallmatrix} \qquad N = \begin{smallmatrix} 3\,1 \\ 4\,2 \\ 3 \\ 4 \end{smallmatrix}.$$

We want to compute the dimension of $\mathrm{Ext}^1(M, N)$.

Our strategy is to use the identity $\mathrm{Ext}^1(M, N) \cong \mathrm{Ext}^1(\tau M, \tau N)$ repeatedly. If N is indecomposable non-projective and such that there is a positive integer t such that $\tau^{t-1} N = P(i)$ is projective (such a module is called *preprojective*) and if M is such that $\tau^s M$ is not projective, for all $s = 0, 1, \ldots, t-1$, then

$$\mathrm{Ext}^1(M, N) \cong \mathrm{Ext}^1(\tau^{t-1} M, P(i)) \cong D\mathrm{Hom}(P(i), \tau^t M) \cong (\tau^t M)_i,$$

where the last identity is given by Theorem 2.11.

In our example, the modules M and N are both non-projective. We can use the Coxeter matrix Φ to compute the dimension vectors of the Auslander–Reiten translations. Since the Cartan matrix is

$$C_A = \begin{bmatrix} 1 & 0 & 0 & 0 \\ 1 & 1 & 0 & 0 \\ 1 & 1 & 1 & 0 \\ 2 & 1 & 1 & 1 \end{bmatrix}$$

the Coxeter matrix $\Phi = -C_A^t C_A^{-1}$ is

$$\Phi = -\begin{bmatrix} 1 & 1 & 1 & 2 \\ 0 & 1 & 1 & 1 \\ 0 & 0 & 1 & 1 \\ 0 & 0 & 0 & 1 \end{bmatrix} \begin{bmatrix} 1 & 0 & 0 & 0 \\ -1 & 1 & 0 & 0 \\ 0 & -1 & 1 & 0 \\ -1 & 0 & -1 & 1 \end{bmatrix} = \begin{bmatrix} 2 & 0 & 1 & -2 \\ 2 & 0 & 0 & -1 \\ 1 & 1 & 0 & -1 \\ 1 & 0 & 1 & -1 \end{bmatrix}.$$

Now $\underline{\dim}\, \tau N = \Phi\, \underline{\dim}\, N = \Phi(1, 1, 2, 2)^t = (0, 0, 0, 1)$, which implies that τN is the simple module $S(4)$ which is projective. On the other hand, $\underline{\dim}\, \tau^2 M = \Phi^2 \underline{\dim}\, M = \Phi^2(1, 0, 0, 1)^t = \Phi(0, 1, 0, 0) = (0, 0, 1, 0)$, so $\tau^2 M = S(3)$ and $\mathrm{Ext}^1(M, N) \cong (\tau^2 M)_4 = 0$.

We end this section with the theorem of existence of almost split sequences

Theorem 7.26. *Let A be a finite-dimensional k-algebra.*

1. *For every indecomposable non-projective A-module M there exists an almost split sequence*

$$0 \longrightarrow \tau M \longrightarrow E \longrightarrow M \longrightarrow 0$$

in mod A.

2. *For every indecomposable non-injective A-module N there exists an almost split sequence*

$$0 \longrightarrow \tau N \longrightarrow E \longrightarrow \tau^{-1} N \longrightarrow 0$$

in mod A.

For a proof see [8, IV.4]. The main idea of the proof is the following. The almost split sequence is a nonzero element of $\mathrm{Ext}^1(M, \tau M)$, and $\mathrm{Ext}^1(M, \tau M)$ is isomorphic to $D\underline{\mathrm{Hom}}(M, M) = D\underline{\mathrm{End}}(M)$, by the Auslander–Reiten formula. Under this isomorphism, the identity $1_M \in \underline{\mathrm{End}}(M)$ corresponds to the almost split sequence in $\mathrm{Ext}^1(M, \tau M)$.

7.5 Notes

The theory of almost split sequences has been developed in [10–13]; for further reading we refer to [8, 15, 52]. The radical of a category has been introduced in [17].

Problems

Exercises for Chap. 6

7.1. Let Q be the quiver

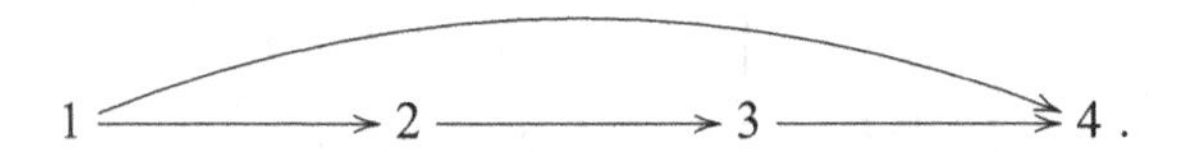

1. Show that $\dim \mathrm{Hom}(P(4), P(1)) = 2$, and conclude that there are two paths in the Auslander–Reiten quiver from $P(4)$ to $P(1)$.
2. Show that the dimension of $\mathrm{rad}\,(P(4), P(1))/\mathrm{rad}^2(P(4), P(1)) = 1$ and conclude that there is only one arrow Auslander–Reiten quiver from $P(4)$ to $P(1)$.
3. Compute the dimension vectors of $\tau^{-1}P(i)$ for $i = 1, 2, 3, 4$ and draw the beginning of the Auslander–Reiten quiver, showing the eight vertices $P(i)$ and $\tau^{-1}P(i)$. You have drawn the beginning of the so-called *preprojective component* of the Auslander–Reiten quiver. This component contains infinitely many indecomposables.
4. Compute the dimension vectors of $\tau I(i)$ for $i = 1, 2, 3, 4$ and draw the ending of the Auslander–Reiten quiver, showing the eight vertices $I(i)$ and $\tau I(i)$. You have drawn the ending of the so-called *preinjective component* of the Auslander–Reiten quiver. This component also contains infinitely many indecomposables.
5. Compute the τ-orbit of $S(3)$ and construct the dimension vectors of at least 9 modules in its component in the Auslander-Reiten quiver. Show that this component contains three distinct indecomposable modules that all have the same dimension vector $(1, 1, 1, 1)$. This component is called a *tube* of rank three and is a so-called *regular component* of the Auslander–Reiten quiver.
6. Show that there are infinitely many distinct A-modules with dimension vector $(1, 1, 1, 1)$. [Hint: Show that each $\lambda \in k$ gives a distinct representations]

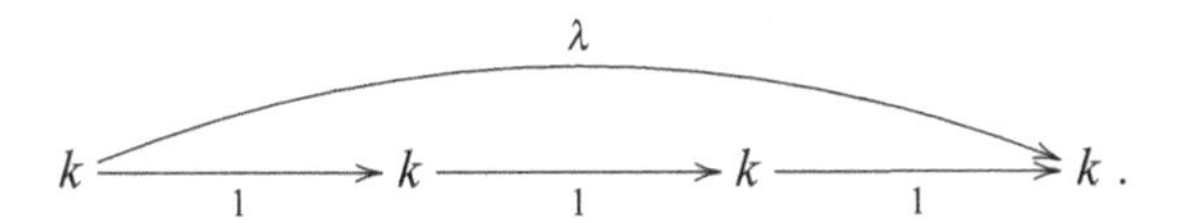

7. Show that $\mathrm{Ext}^1(S(2), P(i)) = \begin{cases} 0 \text{ if } i = 1, 2, 4 \\ k \text{ if } i = 3. \end{cases}$

7.2. Prove that $\mathrm{Ext}^1(S(i), S(j)) \neq 0$ if and only if there is an arrow $j \to i$ in Q.

7.3. Prove that $\mathrm{Ext}^2(S(i), S(j)) \neq 0$ if and only if there is a relation $j \rightsquigarrow i$ in Q.

7.4. Let Q be the quiver

$$1 \xrightarrow{\alpha} 2 \xrightarrow{\beta} 3 \xrightarrow{\gamma} 4 \xrightarrow{\delta} 5$$

bound by the relations $\alpha\beta = 0$ and $\gamma\delta = 0$. Prove that $A = kQ$ has global dimension 2.

7.5. Let Q be the quiver

$$1 \xrightarrow{\alpha} 2 \xrightarrow{\beta} 3 \xrightarrow{\gamma} 4 \xrightarrow{\delta} 5$$

bound by the relations $\alpha\beta\gamma = 0$ and $\beta\gamma\delta = 0$. Prove that $A = kQ$ has global dimension 3.

7.6. Prove that an algebra A is hereditary if and only if the global dimension of A is at most 1.

Chapter 8
Quadratic Forms and Gabriel's Theorem

The main goal of this chapter is to prove that the number of isoclasses of indecomposable representations of a connected quiver Q is finite if and only if Q is of Dynkin type $\mathbb{A}, \mathbb{D}$ or $\mathbb{E}$. The proof we are presenting uses the classification of positive definite integral quadratic forms associated to graphs and also a little algebraic geometry. For a different proof, using tilting theory, see [8, VII.5].

8.1 Variety of Representations

Let $Q = (Q_0, Q_1)$ be a quiver without oriented cycles, and let n be the number of vertices. Let $\mathbf{d} = (d_i) \in \mathbb{Z}^n$ such that $d_i \geq 0$ for all i; we will call $\mathbf{d}$ a *dimension vector*.

Let $E_{\mathbf{d}}$ be the space of all representations $M = (M_i, \varphi_\alpha)_{i \in Q_0, \alpha \in Q_1}$ of Q of dimension vector $\mathbf{d}$; thus $M_i \cong k^{d_i}$, for all $i \in Q_0$. In other words, the vector spaces M_i of the representations in $E_{\mathbf{d}}$ are fixed (up to isomorphism) and the representations in $E_{\mathbf{d}}$ are completely determined by their linear maps φ_α. Thus

$$E_{\mathbf{d}} \cong \bigoplus_{\alpha \in Q_1} \operatorname{Hom}_k(k^{d_{s(\alpha)}}, k^{d_{t(\alpha)}}),$$

and each $\operatorname{Hom}_k(k^{d_{s(\alpha)}}, k^{d_{t(\alpha)}})$ is isomorphic to the space of $d_{t(\alpha)} \times d_{s(\alpha)}$ matrices with entries in k. In particular, $E_{\mathbf{d}}$ is a k-vector space of dimension $\sum_\alpha d_{s(\alpha)} d_{t(\alpha)}$.

Let

$$G_{\mathbf{d}} = \prod_{i \in Q_0} \operatorname{GL}_{d_i}(k),$$

R. Schiffler, *Quiver Representations*, CMS Books in Mathematics,
DOI 10.1007/978-3-319-09204-1_8

where $\mathrm{GL}_{d_i}(k)$ is the group of invertible linear maps from k^{d_i} to itself. The group $G_{\mathbf{d}}$ acts on the space $E_{\mathbf{d}}$ by conjugation; more precisely, if $g = (g_i) \in G_{\mathbf{d}}$, $M = (M_i, \varphi_\alpha) \in E_{\mathbf{d}}$, and $i \stackrel{\alpha}{\to} j$ is an arrow in Q, then $(g \cdot \varphi)_\alpha = g_j \varphi_\alpha g_i^{-1}$:

$$\begin{array}{ccc} k^{d_i} & \xrightarrow{\varphi_\alpha} & k^{d_j} \\ \circlearrowleft & & \circlearrowright \\ g_i & & g_j \end{array}$$

We will denote the orbit of a representation M under this action by $\mathcal{O}_M$; thus $\mathcal{O}_M = \{g \cdot M \mid g \in G_{\mathbf{d}}\}$.

Example 8.1. If Q is the quiver $1 \longrightarrow 2$ and $\mathbf{d} = (d_1, d_2)$, then $E_{\mathbf{d}}$ is isomorphic to the space of $d_2 \times d_1$ matrices, the representation $M \in E_{\mathbf{d}}$ is given by a such a matrix φ, the elements (g_1, g_2) of the group $G_{\mathbf{d}}$ are pairs of invertible square matrices of size d_1 and d_2, and the orbit $\mathcal{O}_M = \{g_2 \varphi g_1^{-1} \mid (g_1, g_2) \in G_{\mathbf{d}}\}$ is the set of all matrices whose rank is equal to the rank of φ. We have seen in Chap. 1 that this set is exactly the isoclass of the representation M. We show in the following lemma that this is not a coincidence:

Lemma 8.1. *The orbit $\mathcal{O}_M$ is precisely the isoclass of the representation M, that is,*

$$\mathcal{O}_M = \{M' \in \operatorname{rep} Q \mid M' \cong M\}.$$

Proof. Suppose $M = (M_i, \varphi_\alpha)$ and $M' = (M'_i, \varphi'_\alpha)$ are in the same orbit; then there exists $g = (g_i)_{i \in Q_0}$ such that $g \cdot M = M'$, which means that for each arrow $i \stackrel{\alpha}{\to} j$ we have a commutative diagram:

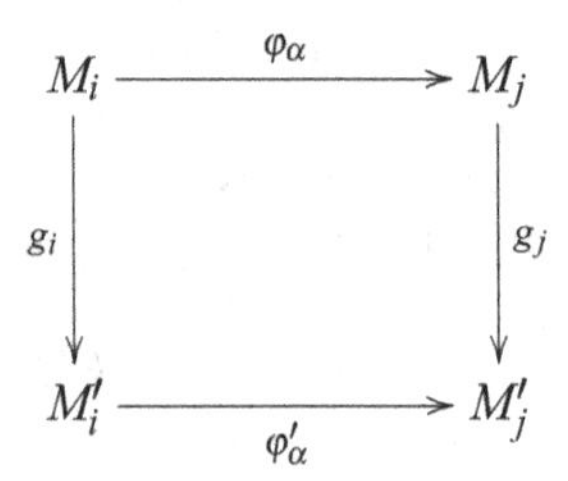

thus g is a morphism of representations. Moreover, each g_i is an element of $\mathrm{GL}_{d_i}(k)$, thus an isomorphism. It follows that M and M' are isomorphic representations.

On the other hand, if $g : M \to M'$ is an isomorphism of representations, then each g_i is an element of $GL_{d_i}(k)$, and hence $M' = g(M) = g \cdot M$. □

The stabilizer $\operatorname{Stab} M = \{g \in G_{\mathbf{d}} \mid g \cdot M = M\}$ of a representation M corresponds to the automorphism group $\operatorname{Aut} M$ of the representation.

We will use the following facts from algebraic geometry:

Lemma 8.2. *Let* $\mathbf{d} \in \mathbb{Z}^n$. *Then*

1. *for any representation* M *of dimension vector* $\mathbf{d}$, *the dimensions of the varieties* $\mathcal{O}_M$, $G_{\mathbf{d}}$, *and* Aut M *satisfy*

$$\dim \mathcal{O}_M = \dim G_{\mathbf{d}} - \dim \operatorname{Aut} M;$$

2. *there is at most one orbit* $\mathcal{O}$ *of codimension zero in* $E_{\mathbf{d}}$.

Proof.

(1) The morphism $G_{\mathbf{d}}/\operatorname{Stab} M \to \mathcal{O}_M$ defined by sending the class $\bar{g}$ of an element $g \in G_{\mathbf{d}}$ to the element $g \cdot M \in \mathcal{O}_M$ is bijective. This shows that $G_{\mathbf{d}}/\operatorname{Stab} M$ and $\mathcal{O}_M$ have the same dimension, and the result then follows from $\operatorname{Stab} M = \operatorname{Aut} M$.

(2) This holds because the algebraic variety $E_{\mathbf{d}}$ is irreducible (i.e., any nonempty open subset is dense), and a codimension zero orbit is open in $E_{\mathbf{d}}$.

□

Lemma 8.3. *If*

$$0 \longrightarrow L \xrightarrow{f} M \xrightarrow{g} N \longrightarrow 0$$

is a non-split short exact sequence of representations, then

$$\dim \mathcal{O}_{L\oplus N} < \dim \mathcal{O}_M.$$

Proof. We use the notation $L = (L_i, \psi_\alpha)$, $M = (M_i, \varphi_\alpha)$ and $N = (N_i, \chi_\alpha)$. For each vertex i, we choose a basis B' of the vector space L_i, extend its image $f_i(B_i)$ under f_i to a basis B of M_i, and choose $B'' = g_i(B)$ as a basis of N_i. With respect to these bases, the morphisms f_i and g_i are given by the following block matrices:

$$f_i = \left[\begin{array}{c} \begin{matrix} 1 & & 0 \\ & \ddots & \\ 0 & & 1 \end{matrix} \\ \hline 0 \end{array}\right] \qquad g_i = \left[\begin{array}{c|c} 0 & \begin{matrix} 1 & & 0 \\ & \ddots & \\ 0 & & 1 \end{matrix} \end{array}\right].$$

Now, let $\alpha : i \to j$ be an arrow in Q. Then the equations $\varphi_\alpha f_i = f_j \psi_\alpha$ and $g_j \varphi_\alpha = \chi_\alpha g_i$ imply that φ_α is represented by a block matrix:

$$\varphi_\alpha = \left[\begin{array}{c|c} \psi_\alpha & \xi_\alpha \\ \hline 0 & \chi_\alpha \end{array}\right],$$

where ξ_α is some matrix of size $(\dim M_j - \dim N_j) \times (\dim M_i - \dim L_i)$. Then M is isomorphic to the direct sum $L \oplus N$ if and only if $\xi_\alpha = 0$, for all α. On the other hand, if M is not isomorphic to $L \oplus N$, then $\xi_\alpha \neq 0$ for some arrow α, and then, for any nonzero $t \in k$, the matrices

$$t \cdot \varphi_\alpha = \left[\begin{array}{c|c} \psi_\alpha & t\xi_\alpha \\ \hline 0 & \chi_\alpha \end{array}\right]$$

define isomorphic representations $t \cdot M = (M_i, t \cdot \varphi_\alpha) \cong M$. An explicit isomorphism is given by the matrices:

$$\left[\begin{array}{c|c} t1_{L_i} & 0 \\ \hline 0 & 1_{N_i} \end{array}\right].$$

This shows that $\dim \mathcal{O}_{L\oplus N} < \dim \mathcal{O}_M$. □

Example 8.2. Let Q be the quiver

$$1 \longrightarrow 2 \longleftarrow 3,$$

and let $\mathbf{d} = (1, 2, 1)$. Recall that the Auslander–Reiten quiver of Q is of the form

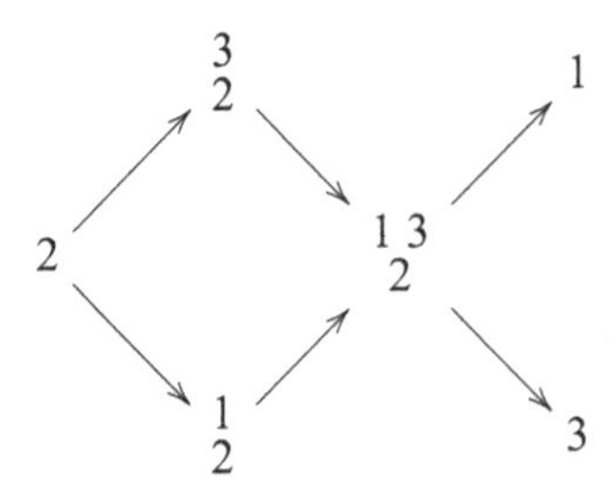

Every representation $M \in E_{\mathbf{d}}$ can be written as a direct sum $M = \oplus c_i M_i$, where M_i is an indecomposable representation and $c_i \geq 0$ denotes its multiplicity. For example, the direct sum of simple representations $\oplus d_i S(i)$ is an element of $E_{\mathbf{d}}$. Since all the maps of the representation $\oplus d_i S(i)$ are zero, it follows that its orbit consists of exactly one point. Hence $\dim \mathcal{O}_{\oplus d_i S(i)} = 0$.

We can visualize each representation by labeling the vertices M_i of the Auslander–Reiten quiver with the multiplicity c_i. For example, the representation

$$S(2) \oplus S(2) \oplus S(2) \oplus \begin{smallmatrix} 1\,3 \\ 2 \end{smallmatrix} \oplus \begin{smallmatrix} 1\,3 \\ 2 \end{smallmatrix} \oplus S(3)$$

corresponds to

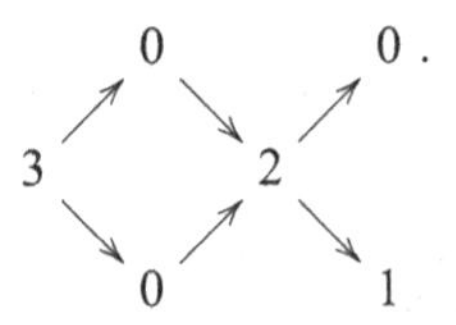

Now, we can use short exact sequences to build representatives of the other orbits in $E_{\mathbf{d}}$. For example, the short exact sequence

$$0 \longrightarrow S(2) \longrightarrow \begin{smallmatrix} 1 \\ 2 \end{smallmatrix} \longrightarrow S(1) \longrightarrow 0$$

gives the representation $\begin{smallmatrix} 1 \\ 2 \end{smallmatrix} \oplus S(2) \oplus S(3)$ obtained from $\oplus d_i S(i)$ by replacing the summand $S(2) \oplus S(1)$ by $\begin{smallmatrix} 1 \\ 2 \end{smallmatrix}$. In our symbolic notation this operation corresponds to

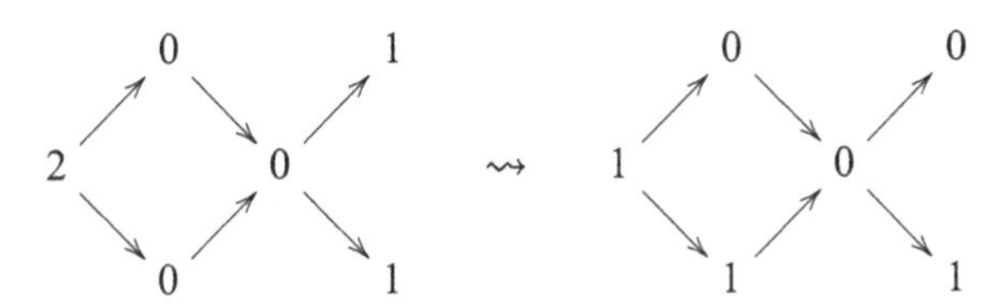

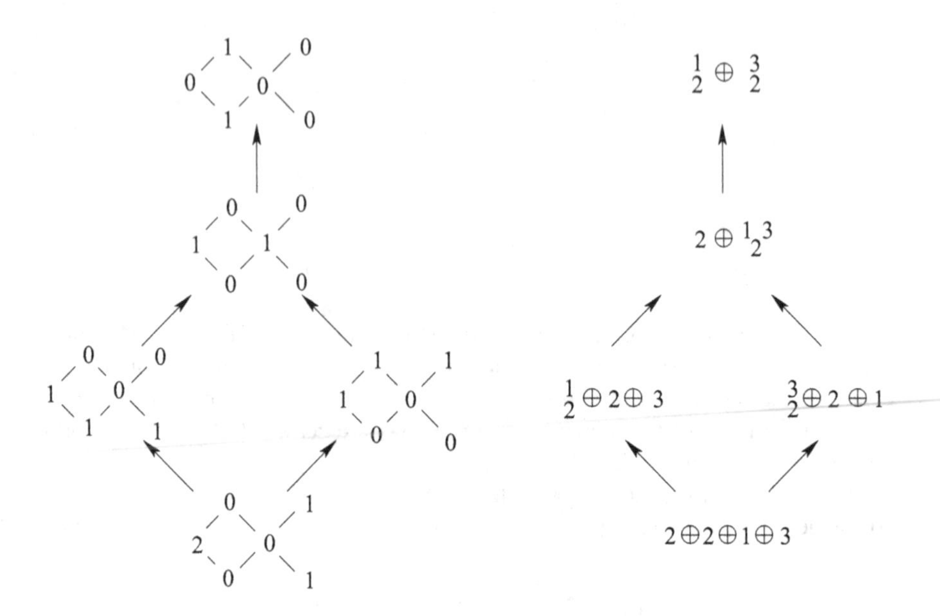

Fig. 8.1 Decomposition of $E_{\mathbf{d}}$ into orbits

because we reduce the multiplicity of $S(2)$ from 2 to 1 and that of $S(1)$ from 1 to 0, and we increase the multiplicity of $\begin{smallmatrix}1\\2\end{smallmatrix}$ from 0 to 1.

Taking into account all such operations, we can construct a decomposition of $E_{\mathbf{d}}$ into orbits as shown in Fig. 8.1. The left picture shows the multiplicities in the direct sum decompositions and the right picture shows the corresponding direct sums. It follows from Lemma 8.3 that the dimensions of the orbits are strictly increasing when we follow the arrows in the figure.

8.2 Quadratic Form of a Quiver

Let Q be a quiver without oriented cycles. In this section we will define a quadratic form q associated to Q and introduce its roots.

Let us start by recalling that an *n-ary integral quadratic form q* is a homogeneous polynomial of degree 2 in n variables $x_1, x_2, \ldots, x_n$ and with coefficients in $\mathbb{Z}$; thus q is of the form

$$q(x_1, x_2, \ldots, x_n) = \sum_{i,j=1}^{n} a_{ij} x_i x_j,$$

for some $a_{ij} \in \mathbb{Z}$.

We will often think of a quadratic form as a map:

$$q: \mathbb{Z}^n \to \mathbb{Z}, \mathbf{x} = (x_1, x_2, \ldots, x_n) \mapsto q(\mathbf{x}).$$

Note that $q(r\mathbf{x}) = r^2 q(\mathbf{x})$ for all $r \in \mathbb{Z}$.

Given a quadratic form q, we can define its symmetric bilinear form $(\, , \,)$ by the formula

$$(\mathbf{x}, \mathbf{y}) = q(\mathbf{x} + \mathbf{y}) - q(\mathbf{x}) - q(\mathbf{y}). \tag{8.1}$$

A simple calculation shows that $(\mathbf{x}, \mathbf{y}) = \sum a_{ij}(x_i y_j + x_j y_i)$. We leave it as an exercise to show that $(\, , \,)$ is bilinear.

Conversely, given the symmetric bilinear form $(\, , \,)$, one can recover the quadratic form by the formula

$$q(\mathbf{x}) = \tfrac{1}{2}(\mathbf{x}, \mathbf{x}).$$

Now let Q be a quiver without oriented cycles and define its quadratic form by

$$q : \mathbb{Z}^n \to \mathbb{Z}, \qquad q(\mathbf{x}) = \sum_{i \in Q_0} x_i^2 - \sum_{\alpha \in Q_1} x_{s(\alpha)} x_{t(\alpha)}.$$

Note that q does not depend on the actual orientation of the arrows in the quiver but only on the underlying graph.

Example 8.3. The quadratic form of the quiver

$$1 \longrightarrow 2 \longleftarrow 3$$

is

$$q(\mathbf{x}) = x_1^2 + x_2^2 + x_3^2 - x_1 x_2 - x_2 x_3.$$

We want to evaluate q on the dimension vectors $\mathbf{d}$ of representations of Q. It is trivial, but important to note that the value of the quadratic form q only depends on the dimension vector of a given representation and not on the particular representation itself. In particular, q is constant on the space $E_{\mathbf{d}}$ introduced in the previous section.

The following proposition provides a different interpretation of q in terms of representation theory:

Proposition 8.4. *For any representation M of dimension vector $\mathbf{d}$ we have $q(\mathbf{d}) =$ $\dim \operatorname{Hom}(M, M) - \dim \operatorname{Ext}^1(M, M)$.*

Proof. Consider the standard projective resolution (see (2.3) in the proof of Theorem 2.15)

$$0 \longrightarrow \bigoplus_{\alpha \in Q_1} d_{s(\alpha)} P(t(\alpha)) \xrightarrow{f} \bigoplus_{i \in Q_0} d_i P(i) \xrightarrow{g} M \longrightarrow 0.$$

Applying the functor $\mathrm{Hom}(-, M)$ to it yields an exact sequence

$$0 \to \mathrm{Hom}(M,M) \to \bigoplus_{i \in Q_0} d_i \mathrm{Hom}(P(i),M) \to \bigoplus_{\alpha \in Q_1} d_{s(\alpha)} \mathrm{Hom}(P(t(\alpha)),M) \to \mathrm{Ext}^1(M,M) \longrightarrow 0,$$

where the last term in this sequence is zero, since each $P(i)$ is projective. We can thus conclude that $\dim \mathrm{Hom}(M, M) - \dim \mathrm{Ext}^1(M, M)$ is equal to

$$\sum_{i \in Q_0} d_i \dim \mathrm{Hom}(P(i), M) - \sum_{\alpha \in Q_1} d_{s(\alpha)} \dim \mathrm{Hom}(P(t(\alpha)), M),$$

which, by Theorem 2.11, is equal to

$$\sum_{i \in Q_0} d_i^2 - \sum_{\alpha \in Q_1} d_{s(\alpha)} d_{t(\alpha)}.$$

□

8.2.1 Classification of Positive Definite Quadratic Forms

In Fig. 3.7 of Chap. 3, we have already seen the Dynkin diagrams. We will need here another list of diagrams, called the *Euclidean diagrams* or *affine Dynkin diagrams*; see Fig. 8.2. These diagrams are obtained by extending the Dynkin diagrams of type $\mathbb{A}, \mathbb{D}$ or $\mathbb{E}$ at one point, that is, by adding one vertex (and certain edges) in such a way that the Euclidean diagram is not a Dynkin diagram itself, but if one deletes any vertex from the Euclidean diagram, one obtains a union of Dynkin diagrams. For this reason the Euclidean diagrams are also called extended Dynkin diagrams. If an Euclidean diagram is obtained by extending a Dynkin diagram of type Δ, then the Euclidean diagram is said to be of type $\tilde{\Delta}$. In Fig. 8.2, the extending vertex is labeled $n+1$. Note that the index in the type for an Euclidean diagram is the number of vertices minus one.

A key step in the proof of Gabriel's theorem is the classification of the quivers Q that give rise to a positive definite quadratic form q.

Definition 8.1. Let q be a quadratic form:

1. q is called **positive definite** if $q(\mathbf{x}) > 0$, for all $\mathbf{x} \neq 0$.

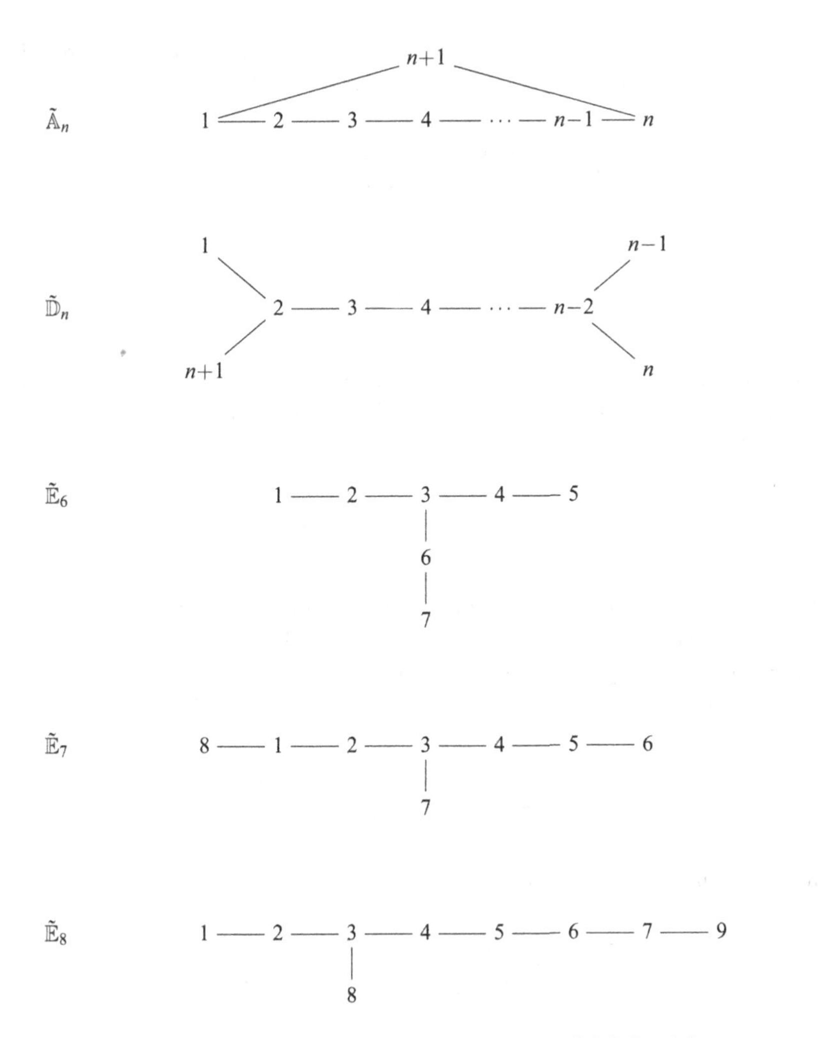

Fig. 8.2 Euclidean diagrams; the extending vertices are labeled $n + 1$

2. q is called **positive semi-definite** if $q(\mathbf{x}) \geq 0$, for all $\mathbf{x}$.

We need the following lemma:

Lemma 8.5. *Assume that Q is connected. Let $\mathbf{d} = (d_i) \in \mathbb{Z}^n \setminus \{0\}$ be such that $(\mathbf{d}, \mathbf{x}) = 0$ for all $\mathbf{x} \in \mathbb{Z}^n$. Then*

1. *q is positive semi-definite.*
2. *$d_i \neq 0$ for all i*
3. *$q(\mathbf{x}) = 0$ if and only if $\mathbf{x} = \frac{a}{b}\mathbf{d}$, for some integers a, b.*

Proof. We will use the notation n_{ij} for the number of arrows from i to j plus the number of arrows from j to i. In other words, n_{ij} is the number of edges in the underlying graph of Q. We also fix an (arbitrary) ordering of the vertices by labeling them by $1, 2, \ldots, n$. With this notation, we have

$$q(\mathbf{x}) = \sum_{i=1}^{n} x_i^2 - \sum_{i=1}^{n} \sum_{j:j<i} n_{ij} x_i x_j \quad \text{and}$$

$$(\mathbf{x}, \mathbf{y}) = \sum_{i=1}^{n} 2x_i y_i - \sum_{i=1}^{n} \sum_{j:j\neq i} n_{ij} x_i y_j .$$

Now suppose that $\mathbf{d}$ satisfies the hypotheses of the lemma, and let $\mathbf{e}_i = (0, \ldots, 0, 1, 0, \ldots, 0)$ be the ith standard basis vector of $\mathbb{Z}^n$. Then $0 = (\mathbf{d}, \mathbf{e}_i) = 2d_i - \sum_{j:j\neq i} n_{ij} d_j$; thus

$$d_i = \sum_{j:j<i} n_{ij} d_j . \tag{8.2}$$

Since $n_{ij} \geq 0$, it follows that if there exists a vertex i such that $d_i = 0$, then for all its neighbors j, we also have $d_j = 0$. Since Q is connected, this implies that $d_j = 0$ for all $j \in Q_0$, a contradiction. This shows (2).

Now let $\mathbf{x} \in \mathbb{Z}^n$. Then, using (8.2), we have

$$\sum_{i=1}^{n} x_i^2 = \sum_{i} \frac{x_i^2}{d_i} \sum_{j:j<i} n_{ij} d_j ,$$

and, rearranging this last expression further, we get

$$\begin{aligned}
\sum_i x_i^2 &= \sum_i \sum_{j:j<i} n_{ij}\, d_j\, \frac{x_i^2}{d_i} \\
&= \sum_i \sum_{j:j\neq i} \frac{n_{ij}\, d_j}{2} \frac{x_i^2}{d_i} && (8.3) \\
&= \sum_i \sum_{j:j<i} \left(\frac{n_{ij}\, d_j}{2} \frac{x_i^2}{d_i} + \frac{n_{ij}\, d_i}{2} \frac{x_j^2}{d_j} \right) && (8.4) \\
&= \sum_i \sum_{j:j<i} \frac{n_{ij}\, d_i\, d_j}{2} \left(\frac{x_i^2}{d_i^2} + \frac{x_j^2}{d_j^2} \right),
\end{aligned}$$

where (8.3) holds since the change in the summation from ordered pairs $j < i$ to pairs $j \neq i$ is compensated by the division by 2 and (8.4) holds since the change in the summation from pairs $j \neq i$ to ordered pairs $j < i$ is compensated by adding the second summand.

Therefore,

$$
\begin{aligned}
q(x) &= \sum_i x_i^2 - \sum_i \sum_{j:j<i} n_{ij}\, x_i\, x_j \\
&= \sum_i \sum_{j:j<i} \frac{n_{ij}\, d_i\, d_j}{2} \left(\frac{x_i^2}{d_i^2} + \frac{x_j^2}{d_j^2} - 2\frac{x_i\, x_j}{d_i\, d_j} \right) \\
&= \sum_i \sum_{j:j<i} \frac{n_{ij}\, d_i\, d_j}{2} \left(\frac{x_i}{d_i} - \frac{x_j}{d_j} \right)^2, \qquad (8.5)
\end{aligned}
$$

which shows that $q(\mathbf{x}) \geq 0$, because $d_i, d_j > 0$ and $n_{ij} \geq 0$. This proves (1). Moreover, (8.5) shows that $q(\mathbf{x}) = 0$ if and only if $\frac{x_i}{d_i} = \frac{x_j}{d_j}$, whenever $n_{ij} \neq 0$. Since Q is connected, it follows that $q(\mathbf{x}) = 0$ if and only if $\frac{x_i}{d_i} = \frac{x_j}{d_j}$, for all vertices $i, j \in Q_0$. This shows (3). □

We are ready for the main theorem of this section.

Theorem 8.6. *Let Q be a connected quiver. Then*

1. *q is positive definite if and only if Q is of Dynkin type $\mathbb{A}, \mathbb{D}, \mathbb{E}$.*
2. *q is positive semi-definite if and only if Q is of Euclidean type $\tilde{\mathbb{A}}, \tilde{\mathbb{D}}, \tilde{\mathbb{E}}$, or of Dynkin type $\mathbb{A}, \mathbb{D}, \mathbb{E}$.*

Proof. We start by showing that q is positive semi-definite if Q is Euclidean. According to Lemma 8.5, it suffices to find a vector δ for each Euclidean diagram such that $(\delta, \mathbf{x}) = 0$ for all $\mathbf{x}$. The vectors δ are shown in Fig. 8.3.

It is easy to check that $(\delta, \mathbf{x}) = \sum_{i=1}^n 2\delta_i x_j - \sum_{i=1}^n \sum_{j:j\neq i} n_{ij} \delta_i x_j = 0$ for all $\mathbf{x}$. For example, in type $\tilde{\mathbb{E}}_6$, we have

$$
\begin{aligned}
(\delta, \mathbf{x}) = {} & 2x_1 + 4x_2 + 6x_3 + 4x_4 + 2x_5 + 4x_6 + 2x_7 \\
& -2x_1 - x_2 - 2x_3 - 3x_4 - 2x_5 - 3x_6 - 2x_7 \\
& -3x_2 - 2x_3 - x_4 - x_6 \\
& -2x_3 \\
= {} & 0.
\end{aligned}
$$

Conversely, suppose that q is positive semi-definite and Q is not Euclidean nor Dynkin. Then Q contains a proper subquiver Q' of Euclidean type. Let q' be the quadratic form of Q', and let δ be the dimension vector given in Fig. 8.3. If Q and Q' have the same set of vertices, then Q has more arrows than Q', and consequently $0 = q'(\delta) > q(\delta)$, a contradiction. On the other hand, if Q has more vertices than Q', chose a vertex i_0 in Q which is connected by an arrow to a vertex j_0 in Q', and define $\mathbf{x}$ by $x_i = 2\delta_i$ for all $i \in Q_0'$, $x_{i_0} = 1$ and $x_j = 0$ for all other vertices j in Q. Then $q(\mathbf{x}) \geq q'(2\delta) + 1 - 2\delta_{j_0} = 1 - 2\delta_{j_0} < 0$, a contradiction. This shows that if q is positive semi-definite, then Q is Euclidean or Dynkin. Moreover, if q is

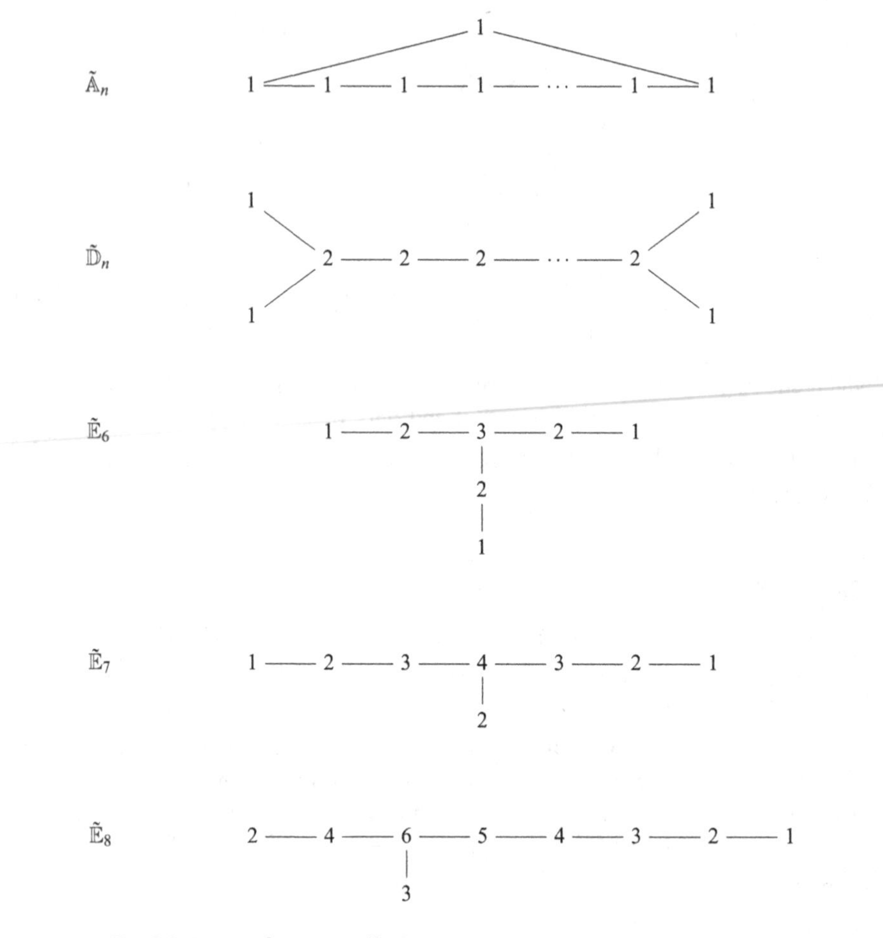

Fig. 8.3 Vectors δ such that $(\delta, \mathbf{x}) = 0$ for all $\mathbf{x}$, for the Euclidean diagrams

positive definite, then q must be Dynkin since for each Euclidean diagram, we have δ with $q(\delta) = 0$.

It only remains to show that, if Q is a Dynkin quiver of type $\mathbb{A}, \mathbb{D}, \mathbb{E}$, then q is positive definite. Let $\overline{Q}$ be the Euclidean quiver obtained by extending Q at one vertex labeled $n+1$, and let $\overline{q}$ denote the quadratic form of $\overline{Q}$.

Suppose there exists $\mathbf{x} \in \mathbb{Z}^n \setminus \{0\}$ such that $q(\mathbf{x}) \leq 0$. Letting $\overline{\mathbf{x}} \in \mathbb{Z}^{n+1}$ be the vector defined by $\overline{x}_i = x_i$ if $i \neq n+1$, and $\overline{x}_{n+1} = 0$, we get $\overline{q}(\overline{\mathbf{x}}) = q(\mathbf{x}) \leq 0$, and therefore $\overline{q}(\overline{\mathbf{x}}) = 0$, since $\overline{q}$ is positive semi-definite. It follows then from Lemma 8.5 that $\overline{\mathbf{x}} = \frac{a}{b}\delta$, for some integers a, b, but this is impossible, since $\overline{x}_{n+1} = 0$. Thus q is positive definite. This completes the proof. □

8.3 Roots

There are two kind of **roots** for a positive semi-definite quadratic form: the real roots and the imaginary roots. An element $\mathbf{x} \in \mathbb{Z}^n \setminus \{0\}$ is called a *real root* if $q(\mathbf{x}) = 1$, and $\mathbf{x}$ is called an *imaginary root* if $q(\mathbf{x}) = 0$. Let Φ denote the set of all roots.

Let $\mathbf{e}_i \in \mathbb{Z}^n$ denote the standard basis vector $(0, \dots, 0, 1, 0, \dots, 0)$ with 1 at the ith position. Of course every element of $\mathbb{Z}^n$, so in particular every root, is of the form $\sum_i a_i \mathbf{e}_i$ with $a_i \in \mathbb{Z}$.

A root $\alpha = \sum_i a_i \mathbf{e}_i$ is called *positive* if all $a_i \geq 0$, and α is called *negative* if all $a_i \leq 0$. Let Φ_+ be the set of positive roots and Φ_- the set of negative roots.

Lemma 8.7.

1. *$\mathbf{e}_i$ is a real root for every $i \in Q_0$.*
2. *If α is a root, then $-\alpha$ is a root.*
3. *If α is a root of an Euclidean quiver and α is not the imaginary root δ of Fig. 8.3, then $\alpha - \delta$ is a root.*
4. *If q is positive semi-definite, then each root is either positive or negative and $\Phi = \Phi_- \sqcup \Phi_+$, and $\Phi_- = -\Phi_+$.*

Remark 8.8. Statement (4) of the lemma also holds when q is not positive semi-definite, but we do not need it here.

Proof. (1) follows directly from the definition of q. (2) holds, because $q(-\alpha) = q((-1)\alpha) = (-1)^2 q(\alpha) = q(\alpha)$. To show (3), we use (8.1) to compute $q(\alpha - \delta) = q(\alpha) + q(-\delta) + (\alpha, -\delta)$, which then implies that $q(\alpha - \delta) = q(\alpha)$, since $q(-\delta) = (\alpha, \delta) = 0$, by definition of δ.

(4) Let $\alpha = \sum_i a_i \mathbf{e}_i$ be a root, and put $\beta = \sum_{i:a_i>0} a_i \mathbf{e}_i$ and $\gamma = \sum_{i:a_i<0} a_i \mathbf{e}_i$; thus $\alpha = \beta + \gamma$. Suppose that β and γ are nonzero, and write

$$\beta = \sum_i \beta_i \mathbf{e}_i, \quad \gamma = \sum_i \gamma_i \mathbf{e}_i.$$

Note that $\beta_i \gamma_i = 0$ and $\beta_i \gamma_j \leq 0$, and therefore $(\beta, \gamma) > 0$. Now since $q(\alpha) = q(\beta + \gamma) = q(\beta) + q(\gamma) + (\beta, \gamma)$, it follows that $q(\alpha) > q(\beta) + q(\gamma)$. But $q(\alpha)$ is either 0 or 1, and thus $q(\beta) + q(\gamma) \leq 0$. Since q is positive semi-definite, it follows that $q(\beta) = q(\gamma) = 0$, and then Lemma 8.5 implies that $\beta = \frac{a}{b}\delta$ for some integers a, b. In particular $\beta_i \neq 0$ for all vertices i. Therefore $\gamma = 0$ and $\alpha = \beta$ is positive. This shows (4). □

Corollary 8.9. *If Q is of Dynkin type, then there are finitely many roots and each root is a real root.*

Proof. There are no imaginary roots since q is positive definite. Let α be a positive root of q. Let $\overline{Q}$ be an Euclidean quiver obtained by extending Q, and denote the new vertex by i_0. Then let $\overline{q}$ be the quadratic form of $\overline{Q}$, and let δ be the imaginary root of $\overline{q}$ given in Fig. 8.3.

Then $\alpha - \delta$ is a root, by Lemma 8.7, which is negative at the vertex i_0, and therefore a negative root, by Lemma 8.7. It follows that for each $i \in Q_0$, $\alpha_i \leq \delta_i$, and there are only finitely many possibilities for α. □

The last proof shows that each positive root of a Dynkin diagram is smaller than the imaginary root of the corresponding Euclidean diagram. This allows us to make a complete list of all positive roots in the Dynkin case.

8.3.1 *Positive Roots in Type* $\mathbb{A}_n$

In type $\mathbb{A}_n$, the positive roots α are smaller than the imaginary root δ in type $\tilde{\mathbb{A}}_n$, thus $\alpha_i \leq 1$ for all $i \in Q_0$. It is easy to check that if α is a positive root, then the set of vertices i such that $\alpha_i = 1$ must form a connected subquiver of Q. Thus the positive roots are indexed by pairs a, b such that $1 \leq a \leq b \leq n$. The number of positive roots is $\frac{n(n+1)}{2}$.

For example, the positive roots in type $\mathbb{A}_3$ are

$$100, 010, 001, 110, 011, 111.$$

8.3.2 *Positive Roots in Type* $\mathbb{D}_n$

In type $\mathbb{D}_n$, the positive roots α are smaller than the imaginary root δ in type $\tilde{\mathbb{D}}_n$, thus $\alpha_i \leq 2$ for all $i \in Q_0$ except for the three outer vertices labeled $1, n-1, n$, on which α is at most 1. Moreover, the type $\mathbb{A}$ subquivers of Q give rise to type $\mathbb{A}$ roots of Q. The positive roots of type $\mathbb{D}$ that do not correspond to subquivers of type $\mathbb{A}$ are the roots

$$\begin{array}{ccccccc} 1 & & 1 & & 1 & & 1 \\ 11\cdots11, & & 11\cdots12, & & 11\cdots122, & \ldots, & 12\cdots22. \\ 1 & & 1 & & 1 & & 1 \end{array}$$

The number of positive roots is $n(n-1)$.

For example the positive roots in type $\mathbb{D}_5$ are

$$\begin{array}{cccccccccc} 0 & 0 & 0 & 1 & 0 & 0 & 0 & 1 & 0 & 0 \\ 100, & 010, & 001, & 000, & 000, & 110, & 011, & 001, & 001, & 111, \\ 0 & 0 & 0 & 0 & 1 & 0 & 0 & 0 & 1 & 0 \end{array}$$

$$\begin{array}{cccccccccc} 1 & 0 & 1 & 1 & 0 & 1 & 1 & 1 & 1 & 1 \\ 011, & 011, & 001, & 111, & 111, & 011, & 111, & 012, & 112, & 122. \\ 0 & 1 & 1 & 0 & 1 & 1 & 1 & 1 & 1 & 1 \end{array}$$

8.3.3 *Positive Roots in Type* $\mathbb{E}_6$

In type $\mathbb{E}_6$, the positive roots α are smaller than the imaginary root δ in type $\tilde{\mathbb{E}}_6$; thus $\alpha \leq \begin{matrix}12321\\2\end{matrix}$. Moreover, the type $\mathbb{A}$ and the type $\mathbb{D}$ subquivers of Q give rise to type $\mathbb{A}$ and type $\mathbb{D}$ roots of Q. There are 36 positive roots; the ones that are not of type $\mathbb{A}$ or $\mathbb{D}$ are

$$\begin{matrix}11111\\1\end{matrix},\ \begin{matrix}11211\\1\end{matrix},\ \begin{matrix}12211\\1\end{matrix},\ \begin{matrix}11221\\1\end{matrix},\ \begin{matrix}12221\\1\end{matrix},\ \begin{matrix}12321\\1\end{matrix},\ \begin{matrix}12321\\2\end{matrix}.$$

8.3.4 *Positive Roots in Type* $\mathbb{E}_7$

In type $\mathbb{E}_7$, the positive roots α are smaller than the imaginary root δ in type $\tilde{\mathbb{E}}_7$; thus $\alpha \leq \begin{matrix}234321\\2\end{matrix}$. There are 63 positive roots; the ones that are not of type $\mathbb{A}, \mathbb{D}$ or $\mathbb{E}_6$ are

$$\begin{matrix}111111\\1\end{matrix},\ \begin{matrix}112111\\1\end{matrix},\ \begin{matrix}112211\\1\end{matrix},\ \begin{matrix}122111\\1\end{matrix},\ \begin{matrix}112221\\1\end{matrix},\ \begin{matrix}122211\\1\end{matrix},\ \begin{matrix}122221\\1\end{matrix},\ \begin{matrix}123211\\1\end{matrix},$$

$$\begin{matrix}123221\\1\end{matrix},\ \begin{matrix}123211\\2\end{matrix},\ \begin{matrix}123321\\1\end{matrix},\ \begin{matrix}123221\\2\end{matrix},\ \begin{matrix}123321\\2\end{matrix},\ \begin{matrix}124321\\2\end{matrix},\ \begin{matrix}134321\\2\end{matrix},\ \begin{matrix}234321\\2\end{matrix}.$$

8.3.5 *Positive Roots in Type* $\mathbb{E}_8$

In type $\mathbb{E}_8$, the positive roots α are smaller than the imaginary root δ in type $\tilde{\mathbb{E}}_8$; thus $\alpha \leq \begin{matrix}2465432\\3\end{matrix}$. There are 120 positive roots; the ones that are not of type $\mathbb{A}, \mathbb{D}$ or $\mathbb{E}_7$ are

$$\begin{matrix}1111111\\1\end{matrix},\ \begin{matrix}1121111\\1\end{matrix},\ \begin{matrix}1221111\\1\end{matrix},\ \begin{matrix}1122111\\1\end{matrix},\ \begin{matrix}1222111\\1\end{matrix},\ \begin{matrix}1122211\\1\end{matrix},\ \begin{matrix}1232111\\1\end{matrix},$$

$$\begin{matrix}1222211\\1\end{matrix},\ \begin{matrix}1122221\\1\end{matrix},\ \begin{matrix}1232111\\2\end{matrix},\ \begin{matrix}1232211\\1\end{matrix},\ \begin{matrix}1222221\\1\end{matrix},\ \begin{matrix}1232211\\2\end{matrix},\ \begin{matrix}1233211\\1\end{matrix},$$

$$\begin{matrix}1232221\\1\end{matrix},\ \begin{matrix}1233211\\2\end{matrix},\ \begin{matrix}1232221\\2\end{matrix},\ \begin{matrix}1233221\\1\end{matrix},\ \begin{matrix}1243211\\2\end{matrix},\ \begin{matrix}1233221\\2\end{matrix},\ \begin{matrix}1233321\\1\end{matrix},$$

$$\begin{matrix}1343211\\2\end{matrix},\ \begin{matrix}1243221\\2\end{matrix},\ \begin{matrix}1233321\\2\end{matrix},\ \begin{matrix}2343211\\2\end{matrix},\ \begin{matrix}1343221\\2\end{matrix},\ \begin{matrix}1243321\\2\end{matrix},\ \begin{matrix}2343221\\2\end{matrix},$$

$$\begin{matrix}1343321\\2\end{matrix},\ \begin{matrix}1244321\\2\end{matrix},\ \begin{matrix}2343321\\2\end{matrix},\ \begin{matrix}1344321\\2\end{matrix},\ \begin{matrix}1354321\\2\end{matrix},\ \begin{matrix}2344321\\2\end{matrix},\ \begin{matrix}1354321\\3\end{matrix},$$

$$\begin{matrix}2354321\\2\end{matrix},\ \begin{matrix}2354321\\3\end{matrix},\ \begin{matrix}2454321\\2\end{matrix},\ \begin{matrix}2454321\\3\end{matrix},\ \begin{matrix}2464321\\3\end{matrix},\ \begin{matrix}2465321\\3\end{matrix},\ \begin{matrix}2465421\\3\end{matrix},$$

$$\begin{matrix}2465431\\3\end{matrix},\ \begin{matrix}2465432\\3\end{matrix}.$$

8.4 Gabriel's Theorem

In this section, we classify the quivers of finite representation type. We start with a result that relates the dimension of the orbit of a representation to the quadratic form.

Proposition 8.10. *Let Q be a connected quiver and let M be a representation of Q of dimension vector* $\mathbf{d}$. *Then*

$$\operatorname{codim} \mathcal{O}_M = \dim \operatorname{End} M - q(\mathbf{d}) = \dim \operatorname{Ext}^1(M, M).$$

Proof. We have $\dim \mathcal{O}_M = \dim G_{\mathbf{d}} - \dim \operatorname{Aut}(M)$, by Lemma 8.2. The group of automorphisms of M is an open subgroup of the group of endomorphisms; thus $\dim \operatorname{Aut}(M) = \dim \operatorname{End}(M)$. On the other hand, each GL_{d_i} is of dimension d_i^2; thus $\dim G_{\mathbf{d}} = \sum_{i \in Q_0} d_i^2$. Therefore

$$\begin{aligned}\operatorname{codim} \mathcal{O}_M &= \dim E_{\mathbf{d}} - \dim \mathcal{O}_M \\ &= \textstyle\sum_{\alpha \in Q_1} d_{s(\alpha)} d_{t(\alpha)} - \sum_{i \in Q_0} d_i^2 + \dim \operatorname{End}(M).\end{aligned}$$

This shows the first equation of the proposition. The second equation follows from Proposition 8.4. □

Corollary 8.11. *If $q(\mathbf{d}) \leq 0$ then there are infinitely many isoclasses of representations of Q of dimension vector* $\mathbf{d}$.

Proof. Let $\mathbf{d}$ be such that $q(\mathbf{d}) \leq 0$ and let M be a representation with $\underline{\dim}\, M = \mathbf{d}$. Then $\operatorname{codim} \mathcal{O}_M \geq \dim \operatorname{End} M \geq 1$, by Proposition 8.10. Thus the dimension of $E_{\mathbf{d}}$

is strictly larger than the dimension of any orbit $\mathscr{O}_M$, which implies that the number of orbits is infinite. □

Theorem 8.12 (Gabriel's Theorem). *Let Q be connected quiver. Then*

1. *Q is of finite representation type if and only if Q is of Dynkin type $\mathbb{A}, \mathbb{D}$ or $\mathbb{E}$.*

2. *If Q is of Dynkin type $\mathbb{A}, \mathbb{D}$ or $\mathbb{E}$, then the dimension vector induces a bijection ψ from isoclasses of indecomposable representations of Q to the set of positive roots:*

$$\psi : \operatorname{ind} Q \longrightarrow \Phi_+ \qquad \psi(M) = \underline{\dim}\, M.$$

Proof. We will show (2) first. *ψ is well defined:* Let M be an indecomposable representation. We must show that $q(\underline{\dim}\, M) = 1$. By Proposition 8.4, it suffices to show that $\operatorname{End} M \cong k$ and that $\dim \operatorname{Ext}^1(M, M) = 0$.

In order to show that $\operatorname{End} M \cong k$, we proceed by induction on the dimension of M. If M is a simple representation, the result is clear. Suppose now that the dimension of M is bigger than 1 and that $\operatorname{End} L \cong k$ for all proper indecomposable subrepresentations L of M.

Suppose that $\operatorname{End} M \ncong k$. Since M is indecomposable, it follows from Corollary 4.20 that every endomorphism of M is of the form $\lambda 1_M + g$ for some $\lambda \in k$ and some nilpotent endomorphism g. Then, since $\operatorname{End} M \ncong k$, there exists a nonzero nilpotent endomorphism $g \in \operatorname{End} M$. Thus $g^m = 0$, for some $m \geq 2$. We may suppose without loss of generality that $m = 2$; otherwise, it suffices to replace g by the endomorphism g^{m-1}. Moreover, among all nonzero endomorphisms whose square is zero, we choose g such that the image of g is of minimal dimension.

Since $g^2 = 0$, we have $\operatorname{im} g \subset \ker g$, and hence there exists an indecomposable summand L of $\ker g$ such that $\operatorname{im} g \cap L$ is nonzero. Let $\pi : \ker g \to L$ be the projection, and let i be the nonzero morphism given by the composition of π and the inclusion $\operatorname{im} g \subset \ker g$:

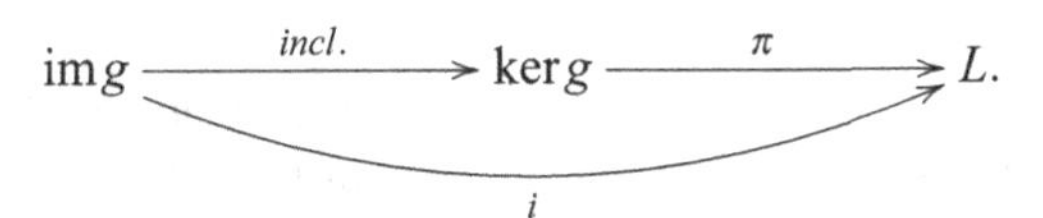

Then the composition

$$M \xrightarrow{\;g\;} \operatorname{im} g \xrightarrow{\;i\;} L \xrightarrow{\;incl.\;} M$$

is a nonzero endomorphism of M whose square is zero and whose image is equal to $i(\operatorname{im} g)$. It then follows from the minimality of g that $\dim i(\operatorname{im} g) \geq \dim \operatorname{im} g$; hence i is injective. We thus get a short exact sequence

$$0 \longrightarrow \operatorname{im} g \xrightarrow{\ i\ } L \longrightarrow \operatorname{coker} i \longrightarrow 0,$$

and applying the functor $\operatorname{Hom}(-, L)$ yields a surjective morphism

$$\operatorname{Ext}^1(L,L) \longrightarrow \operatorname{Ext}^1(\operatorname{im} g, L) \longrightarrow 0. \tag{8.6}$$

By induction, we know that $\dim \operatorname{Hom}(L, L) = 1$, and, since q is positive definite, Proposition 8.4 then implies that $\dim \operatorname{Ext}^1(L, L) = 0$. Therefore, the diagram (8.6) shows that $\operatorname{Ext}^1(\operatorname{im} g, L) = 0$.

Now consider the following commutative diagram with exact rows, whose bottom row is obtained as the push out of the top row along the morphism π, as in Exercise 1.9 of Chap. 1:

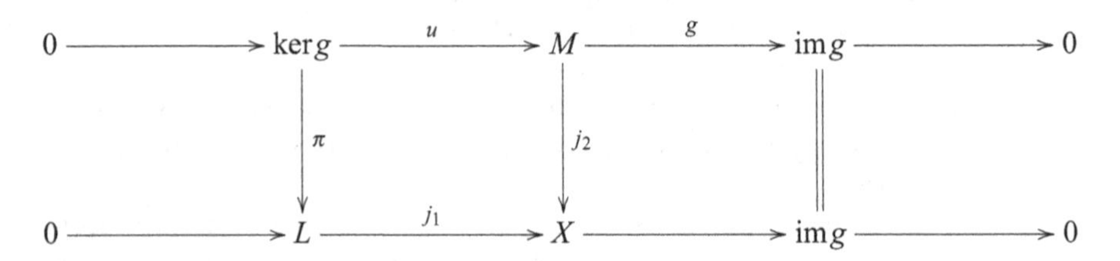

Since $\operatorname{Ext}^1(\operatorname{im} g, L) = 0$, the bottom row splits, so there exists a morphism $h: X \to L$ such that $hj_1 = 1_L$. Let $v: L \to \ker g$ be the inclusion of the direct summand; then we also have $\pi v = 1_L$. Then we can construct morphisms $hj_2: M \to L$ and $uv: L \to M$ with the property that $hj_2uv = hj_1\pi v = 1_L 1_L = 1_L$. This shows that L is a direct summand of the indecomposable representation M; thus $L = 0$ or $L = M$. But we know that $L \neq 0$, because $\operatorname{im} g \cap L$ is nonzero, and we also know that $L \neq M$, because $L \subset \ker g$, and g is nonzero, a contradiction.

This shows that $\dim \operatorname{End} M = 1$. Since q is positive definite, it follows immediately that $\dim \operatorname{Ext}^1(M, M) = 0$ and $q(\underline{\dim}\, M) = 1$. Hence $\underline{\dim}\, M$ is a positive root and ψ is well defined.

ψ is injective: Let M, M' be indecomposable representations such that $\underline{\dim}\, M = \underline{\dim}\, M'$. We have shown above that in Dynkin type the indecomposable representations have no self-extensions. Thus the orbits $\mathcal{O}_M$ and $\mathcal{O}_{M'}$ are both of codimension zero; whence $M \cong M'$, by Lemma 8.2. This shows that ψ is injective.

ψ is surjective: Let Q be Dynkin, let $\mathbf{d}$ a positive root, and let M be a representation, such that $\underline{\dim}\, M = \mathbf{d}$ and $\mathcal{O}_M$ is of maximal dimension in $E_{\mathbf{d}}$. We want to show that M is indecomposable.

Suppose that $M = M_1 \oplus M_2$. The first step is to show that $\operatorname{Ext}^1(M_1, M_2) = \operatorname{Ext}^1(M_2, M_1) = 0$. Suppose that $\operatorname{Ext}^1(M_1, M_2) \neq 0$, then there would be a non-split short exact sequence of the form

$$0 \longrightarrow M_2 \longrightarrow E \longrightarrow M_1 \longrightarrow 0\,.$$

Note that $\underline{\dim}\, E = \underline{\dim}\, M$. Then Lemma 8.3 implies that $\dim \mathscr{O}_M < \dim \mathscr{O}_E$, which contradicts the maximality of $\mathscr{O}_M$. We have shown that $\operatorname{Ext}^1(M_1, M_2) = 0$, and, by symmetry, we also get $\operatorname{Ext}^1(M_2, M_1) = 0$.

Now it follows from Proposition 8.4 that

$$1 = q(\mathbf{d}) = \dim \operatorname{Hom}(M_1 \oplus M_2, M_1 \oplus M_2) \geq 2,$$

a contradiction. Thus M is indecomposable, and $\psi(M) = \mathbf{d}$; whence ψ is surjective. This shows (2).

(1) Suppose that Q is not Dynkin. Then there exists a dimension vector $\mathbf{d} \neq 0$ such that $q(\mathbf{d}) \leq 0$. By Corollary 8.11, there are infinitely many isoclasses of representations of dimension vector $\mathbf{d}$. Each of these representations is a finite direct sum of indecomposable representations, and therefore the number of isoclasses of indecomposable representations is infinite. □

8.5 Notes

Gabriel's theorem was proved in [33] and later in [18]. For alternative proofs of the theorem, see, for example, [8, 15, 52]. Important generalizations were given in [30,43]. Further information on the geometry of quiver representations can be found in [21].

Problems

Exercises for Chap. 8

8.1. Construct the decomposition of $E_{\mathbf{d}}$ into orbits as in Example 8.2 for

1. $Q = 1 \longrightarrow 2 \longleftarrow 3$ and $\mathbf{d} = (2,3,1)$,
2. $Q = 1 \longrightarrow 2 \longrightarrow 3$ and $\mathbf{d} = (2,3,1)$,

8.2. Construct the decomposition of $E_{\mathbf{d}}$ into orbits as in Example 8.2 for the quiver

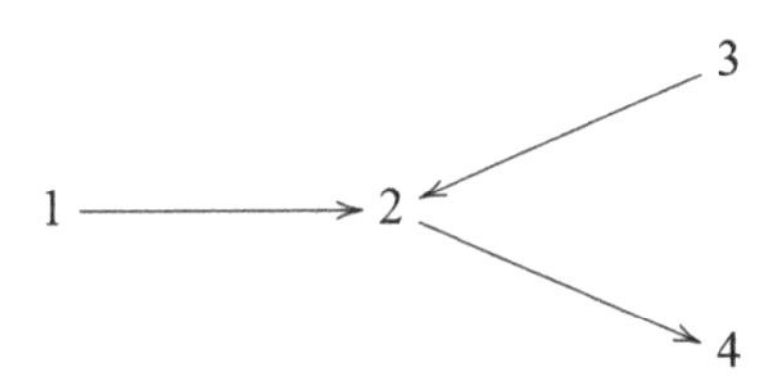

and the dimension vector:

1. $\mathbf{d} = (1, 2, 1, 1)$,
2. $\mathbf{d} = (2, 2, 1, 1)$.

8.3. Let $q : \mathbf{Z}^n \to \mathbf{Z}$ be a quadratic form. Show that the form

$$(\mathbf{x}, \mathbf{y}) = q(\mathbf{x} + \mathbf{y}) - q(\mathbf{x}) - q(\mathbf{y})$$

is symmetric and bilinear.

Exercises 8.2–8.4 use the following terminology: A quiver Q is called a *tree* if its underlying graph has no cycles. Note that parallel edges are cycles. A quiver without oriented cycles which is neither Dynkin nor Euclidean is called *wild*. The *valence* of a vertex in a quiver or graph is the number of its neighbor vertices.

8.4. Prove that for $n \leq 5$ any tree quiver is Dynkin or Euclidean.

8.5. Show that the smallest wild tree with valence at most 3 at each vertex has 7 vertices.

8.6. Show that the smallest wild tree that has at most one vertex of valence 3 has 8 vertices.

8.7. Prove that every quiver that is not Dynkin or Euclidean contains a subquiver that is Euclidean.

8.8. Prove that for every dimension vector $\mathbf{d}$ the space $E_{\mathbf{d}}$ contains an orbit $\mathcal{O}_S$ such that

1. $\mathcal{O}_S$ consists of precisely one representation S.
2. S is semisimple, that is, S is a direct sum of simple representations.
3. $\dim \operatorname{Ext}^1(S, S) = \sum_{\alpha \in Q_1} d_{s(\alpha)} d_{t(\alpha)} = \dim E_{\mathbf{d}}$.

References

1. Frank W. Anderson and Kent R. Fuller, *Rings and categories of modules*, second ed., Graduate Texts in Mathematics, vol. 13, Springer-Verlag, New York, 1992. MR 1245487 (94i:16001)
2. Lidia Angeleri Hügel, Dieter Happel, and Henning Krause (eds.), *Handbook of tilting theory*, London Mathematical Society Lecture Note Series, vol. 332, Cambridge University Press, Cambridge, 2007. MR 2385175 (2008i:16001)
3. Ibrahim Assem, Thomas Brüstle, and Ralf Schiffler, *Cluster-tilted algebras and slices*, J. Algebra **319** (2008), no. 8, 3464–3479. MR 2408327 (2009f:16021)
4. Ibrahim Assem, Thomas Brüstle, and Ralf Schiffler, *Cluster-tilted algebras as trivial extensions*, Bull. Lond. Math. Soc. **40** (2008), no. 1, 151–162. MR 2409188 (2009c:16086)
5. Ibrahim Assem, Thomas Brüstle, and Ralf Schiffler, *On the Galois coverings of a cluster-tilted algebra*, J. Pure Appl. Algebra **213** (2009), no. 7, 1450–1463. MR 2497589 (2010c:16020)
6. Ibrahim Assem, Thomas Brüstle, and Ralf Schiffler, *Cluster-tilted algebras without clusters*, J. Algebra **324** (2010), no. 9, 2475–2502. MR 2684150
7. Ibrahim Assem, Dieter Happel, and Oscar Roldán, *Representation-finite trivial extension algebras*, J. Pure Appl. Algebra **33** (1984), no. 3, 235–242. MR 761629 (85m:16009)
8. Ibrahim Assem, Daniel Simson, and Andrzej Skowroński, *Elements of the representation theory of associative algebras. Vol. 1*, London Mathematical Society Student Texts, vol. 65, Cambridge University Press, Cambridge, 2006, Techniques of representation theory. MR 2197389 (2006j:16020)
9. Maurice Auslander, María Inés Platzeck, and Idun Reiten, *Coxeter functors without diagrams*, Trans. Amer. Math. Soc. **250** (1979), 1–46. MR 530043 (80c:16027)
10. Maurice Auslander and Idun Reiten, *Representation theory of Artin algebras. III. Almost split sequences*, Comm. Algebra **3** (1975), 239–294. MR 0379599 (52 #504)
11. Maurice Auslander and Idun Reiten, *Representation theory of Artin algebras. IV. Invariants given by almost split sequences*, Comm. Algebra **5** (1977), no. 5, 443–518. MR 0439881 (55 #12762)
12. Maurice Auslander and Idun Reiten, *Representation theory of Artin algebras. V. Methods for computing almost split sequences and irreducible morphisms*, Comm. Algebra **5** (1977), no. 5, 519–554. MR 0439882 (55 #12763)
13. Maurice Auslander and Idun Reiten, *Representation theory of Artin algebras. VI. A functorial approach to almost split sequences*, Comm. Algebra **6** (1978), no. 3, 257–300. MR 0472919 (57 #12601)
14. Maurice Auslander, Idun Reiten, and Smalø Sverre O., *Representation theory of Artin algebras*, Cambridge Studies in Advanced Mathematics, vol. 36, Cambridge University Press, Cambridge, 1997, Corrected reprint of the 1995 original. MR 1476671 (98e:16011)

R. Schiffler, *Quiver Representations*, CMS Books in Mathematics,
DOI 10.1007/978-3-319-09204-1

15. Maurice Auslander, Idun Reiten, and SmaløSverre O., *Representation theory of Artin algebras*, Cambridge Studies in Advanced Mathematics, vol. 36, Cambridge University Press, Cambridge, 1995. MR 1314422 (96c:16015)
16. Michael Barot, Elsa Fernández, María Inés Platzeck, Nilda Isabel Pratti, and Sonia Trepode, *From iterated tilted algebras to cluster-tilted algebras*, Adv. Math. **223** (2010), no. 4, 1468–1494. MR 2581376
17. Raymundo Bautista, *Irreducible morphisms and the radical of a category*, An. Inst. Mat. Univ. Nac. Autónoma México **22** (1982), 83–135 (1983). MR 736555 (86g:16041)
18. I. N. Bernšteĭn, I. M. Gel'fand, and V. A. Ponomarev, *Coxeter functors, and Gabriel's theorem*, Uspehi Mat. Nauk **28** (1973), no. 2(170), 19–33. MR 0393065 (52 #13876)
19. Marco Angel Bertani-Økland, Steffen Oppermann, and Anette Wrålsen, *Constructing tilted algebras from cluster-tilted algebras*, J. Algebra **323** (2010), no. 9, 2408–2428. MR 2602387
20. Grzegorz Bobiński and Aśլak Bakke Buan, *The algebras derived equivalent to gentle cluster tilted algebras*, J. Algebra Appl. **11** (2012), no. 1, 1250012, 26. MR 2900882
21. Klaus Bongartz, *Some geometric aspects of representation theory*, Algebras and modules, I (Trondheim, 1996), CMS Conf. Proc., vol. 23, Amer. Math. Soc., Providence, RI, 1998, pp. 1–27. MR 1648601 (99j:16005)
22. Sheila Brenner and M. C. R. Butler, *Generalizations of the Bernstein-Gel'fand-Ponomarev reflection functors*, Representation theory, II (Proc. Second Internat. Conf., Carleton Univ., Ottawa, Ont., 1979), Lecture Notes in Math., vol. 832, Springer, Berlin, 1980, pp. 103–169. MR 607151 (83e:16031)
23. Aslak Bakke Buan, Robert Marsh, Markus Reineke, Idun Reiten, and Gordana Todorov, *Tilting theory and cluster combinatorics*, Adv. Math. **204** (2006), no. 2, 572–618. MR 2249625 (2007f:16033)
24. Aslak Bakke Buan, Robert J. Marsh, and Idun Reiten, *Cluster-tilted algebras of finite representation type*, J. Algebra **306** (2006), no. 2, 412–431. MR 2271343 (2008f:16032)
25. Aslak Bakke Buan, Robert J. Marsh, and Idun Reiten, *Cluster-tilted algebras*, Trans. Amer. Math. Soc. **359** (2007), no. 1, 323–332 (electronic). MR 2247893 (2007f:16035)
26. Aslak Bakke Buan, Robert J. Marsh, and Idun Reiten, *Cluster mutation via quiver representations*, Comment. Math. Helv. **83** (2008), no. 1, 143–177. MR 2365411 (2008k:16026)
27. Philippe Caldero, Frédéric Chapoton, and Ralf Schiffler, *Quivers with relations and cluster tilted algebras*, Algebr. Represent. Theory **9** (2006), no. 4, 359–376. MR 2250652 (2007f:16036)
28. Philippe Caldero, Frédéric Chapoton, and Ralf Schiffler, *Quivers with relations arising from clusters (A_n case)*, Trans. Amer. Math. Soc. **358** (2006), no. 3, 1347–1364. MR 2187656 (2007a:16025)
29. Edward Cline, Brian Parshall, and Leonard Scott, *Derived categories and Morita theory*, J. Algebra **104** (1986), no. 2, 397–409. MR 866784 (88a:16075)
30. Vlastimil Dlab and Claus Michael Ringel, *Representations of graphs and algebras*, Department of Mathematics, Carleton University, Ottawa, Ont., 1974, Carleton Mathematical Lecture Notes, No. 8. MR 0387350 (52 #8193)
31. David S. Dummit and Richard M. Foote, *Abstract algebra*, third ed., John Wiley & Sons Inc., Hoboken, NJ, 2004. MR 2286236 (2007h:00003)
32. Robert M. Fossum, Phillip A. Griffith, and Idun Reiten, *Trivial extensions of abelian categories*, Lecture Notes in Mathematics, Vol. 456, Springer-Verlag, Berlin, 1975, Homological algebra of trivial extensions of abelian categories with applications to ring theory. MR 0389981 (52 #10810)
33. Peter Gabriel, *Unzerlegbare Darstellungen. I*, Manuscripta Math. **6** (1972), 71–103; correction, ibid. 6 (1972), 309. MR 0332887 (48 #11212)
34. Peter Gabriel, *Indecomposable representations. II*, Symposia Mathematica, Vol. XI (Convegno di Algebra Commutativa, INDAM, Rome, 1971), Academic Press, London, 1973, pp. 81–104. MR 0340377 (49 #5132)

35. Peter Gabriel, *Auslander-Reiten sequences and representation-finite algebras*, Representation theory, I (Proc. Workshop, Carleton Univ., Ottawa, Ont., 1979), Lecture Notes in Math., vol. 831, Springer, Berlin, 1980, pp. 1–71. MR 607140 (82i:16030)
36. Dieter Happel, *On the derived category of a finite-dimensional algebra*, Comment. Math. Helv. **62** (1987), no. 3, 339–389. MR 910167 (89c:16029)
37. Dieter Happel, *A characterization of hereditary categories with tilting object*, Invent. Math. **144** (2001), no. 2, 381–398. MR 1827736 (2002a:18014)
38. Dieter Happel, Idun Reiten, and Smalø Sverre O., *Tilting in abelian categories and quasitilted algebras*, Mem. Amer. Math. Soc. **120** (1996), no. 575, viii+ 88. MR 1327209 (97j:16009)
39. Dieter Happel and Claus Michael Ringel, *Tilted algebras*, Trans. Amer. Math. Soc. **274** (1982), no. 2, 399–443. MR 675063 (84d:16027)
40. Mitsuo Hoshino, *Trivial extensions of tilted algebras*, Comm. Algebra **10** (1982), no. 18, 1965–1999. MR 674704 (84j:16019)
41. David Hughes and Josef Waschbüsch, *Trivial extensions of tilted algebras*, Proc. London Math. Soc. (3) **46** (1983), no. 2, 347–364. MR 693045 (84m:16023)
42. Yasuo Iwanaga and Takayoshi Wakamatsu, *Trivial extension of Artin algebras*, Representation theory, II (Proc. Second Internat. Conf., Carleton Univ., Ottawa, Ont., 1979), Lecture Notes in Math., vol. 832, Springer, Berlin, 1980, pp. 295–301. MR 607160 (82c:16024)
43. V. G. Kac, *Infinite root systems, representations of graphs and invariant theory*, Invent. Math. **56** (1980), no. 1, 57–92. MR 557581 (82j:16050)
44. Bernhard Keller and Idun Reiten, *Cluster-tilted algebras are Gorenstein and stably Calabi-Yau*, Adv. Math. **211** (2007), no. 1, 123–151. MR 2313531 (2008b:18018)
45. T. Y. Lam, *Lectures on modules and rings*, Graduate Texts in Mathematics, vol. 189, Springer-Verlag, New York, 1999. MR 1653294 (99i:16001)
46. T. Y. Lam, *A first course in noncommutative rings*, second ed., Graduate Texts in Mathematics, vol. 131, Springer-Verlag, New York, 2001. MR 1838439 (2002c:16001)
47. Joachim Lambek, *Lectures on rings and modules*, second ed., Chelsea Publishing Co., New York, 1976. MR 0419493 (54 #7514)
48. Miki Oryu and Ralf Schiffler, *On one-point extensions of cluster-tilted algebras*, J. Algebra **357** (2012), 168–182. MR 2905247
49. Richard S Pierce, *Associative algebras*, Springer-Verlag, New York, 1982.
50. Marju Purin, *τ-complexity of cluster tilted algebras*, J. Pure Appl. Algebra **216** (2012), no. 4, 897–904. MR 2864863 (2012k:16033)
51. Jeremy Rickard, *Morita theory for derived categories*, J. London Math. Soc. (2) **39** (1989), no. 3, 436–456. MR 1002456 (91b:18012)
52. Claus Michael Ringel, *Tame algebras and integral quadratic forms*, Lecture Notes in Mathematics, vol. 1099, Springer-Verlag, Berlin, 1984. MR 774589 (87f:16027)
53. Joseph J Rotman, *An introduction to homological algebra*, Springer, 2009.
54. Ralf Schiffler, *A geometric model for cluster categories of type D_n*, J. Algebraic Combin. **27** (2008), no. 1, 1–21. MR 2366159 (2008k:16025)
55. Daniel Simson and Andrzej Skowroński, *Elements of the representation theory of associative algebras. Vol. 2*, London Mathematical Society Student Texts, vol. 71, Cambridge University Press, Cambridge, 2007, Tubes and concealed algebras of Euclidean type. MR 2360503 (2009f:16001)
56. Daniel Simson and Andrzej Skowroński, *Elements of the representation theory of associative algebras. Vol. 3*, London Mathematical Society Student Texts, vol. 72, Cambridge University Press, Cambridge, 2007, Representation-infinite tilted algebras. MR 2382332 (2008m:16001)
57. Hiroyuki Tachikawa, *Representations of trivial extensions of hereditary algebras*, Representation theory, II (Proc. Second Internat. Conf., Carleton Univ., Ottawa, Ont., 1979), Lecture Notes in Math., vol. 832, Springer, Berlin, 1980, pp. 579–599. MR 607173 (82d:16029)
58. Bin Zhu, *Cluster-tilted algebras and their intermediate coverings*, Comm. Algebra **39** (2011), no. 7, 2437–2448. MR 2821722 (2012f:16043)

Index

R. Schiffler, *Quiver Representations*, CMS Books in Mathematics,
DOI 10.1007/978-3-319-09204-1

Printed in the United States
By Bookmasters